Social Forces and the Revolution in Military Affairs

Social Forces and the Revolution in Military Affairs

The Cases of Colombia and Mexico

James F. Rochlin

palgrave
macmillan

SOCIAL FORCES AND THE REVOLUTION IN MILITARY AFFAIRS

First published in 2007 by
PALGRAVE MACMILLAN™
175 Fifth Avenue, New York, N.Y. 10010 and
Houndmills, Basingstoke, Hampshire, England RG21 6XS.
Companies and representatives throughout the world.

PALGRAVE MACMILLAN is the global academic imprint of the Palgrave Macmillan division of St. Martin's Press, LLC and of Palgrave Macmillan Ltd. Macmillan® is a registered trademark in the United States, United Kingdom and other countries. Palgrave is a registered trademark in the European Union and other countries.

ISBN-13: 978-0-230-60282-3
ISBN-10: 0-230-60282-7

Library of Congress Cataloging-in-Publication Data

Rochlin, James Francis, 1956–
Social forces and the revolution in military affairs : the cases of Colombia and Mexico / James F. Rochlin.
p. cm.
Includes bibliographical references and index.
ISBN 0-230-60282-7
1. Colombia—Military policy. 2. Military planning—Colombia. 3. Mexico—Military policy. 4. Military planning—Mexico. 5. War—Effect of technology on. 6. Low-intensity conflicts (Military science). 7. Social problems—Colombia. 8. Social problems—Mexico. 9. Military art and science—Technological innovations. 10. Insurgency—Colombia. 11. Insurgency—Mexico. I. Title.
UA625.R62 2007
355'.033072—dc22 2007021997

A catalogue record of the book is available from the British Library.

Design by Westchester Book Group

First edition: December 2007

10 9 8 7 6 5 4 3 2 1

Printed in the United States of America.

Contents

List of Tables

Acknowledgments

I would especially like to thank Dr. Bruce Bagley for his helpful encouragement and crucial support. I am deeply grateful for the generous research grants from the Social Sciences and Humanities Research Council of Canada, and for the outstanding research assistance of Ryan MacKay, Barbara Sobol, and Laura Mandelbaum. I am also grateful to the faculty members of the Unidad Academica de Filosofia y Letras of the Universidad Autonoma de Guerrero, the Departamento de Ciencia Politica at the Universidad de los Andes, and the Instituto de Estudios Políticos y Relaciones Internacionales at the Universidad Nacional, Bogotá, for their inspiration. I am grateful to the University of British Columbia for the research assistance offered in many areas. I especially want to thank Ruth for her loving support.

Acronyms

Area de Libre Comercio de las Americas (ALCA)
Asamblea Popular de los Pueblos de Oaxaca (APPO)
Asociación de AfroColombianos Desplazados (AFRODES)
Autodefensas Unidas de Colombia (AUC)
Central American Free Trade Agreement (Cafta)
Central Unitaria de Trabajadores de Colombia (CUT)
Centro de Derechos Humanos Tlachinollan (CDHT)
Centro de Información de Seguridad Nacional (CISEN)
Confederación de Trabajadores de Colombia (CTC)
Departamento Administrativo de Seguridad (DAS)
Ejército de Liberación Nacional (ELN)
Ejército Popular Revolucionario (EPR)
Ejército Revolucionario del Pueblo Insurgente (EPRI)
Ejército Zapatista de Liberación Nacional (EZLN)
Fuerzas Armadas Revolucionarias de Colombia (FARC)
Fuerzas Armadas Revolucionarias del Pueblo (FARP)
International Monetary Fund (IMF)
Military Professional Resources Incorporated (MPRI)
Muerte a Secuestradores (MAS)
nongovernmental organizations (NGOs)
North American Free Trade Agreement (Nafta)
Organización de Campesinos de Sierra del Sur (OCSS)
Partido de la Revolución Democrática (PRD)
Partido de los Pobres (PDLP)
Partido Revolucionario Institucional (PRI)
Partido Revolucionario Obrero Clandestino Union del Pueblo (PROCUP)
Plan Colombia (PC)
Plan Puebla Panama (PPP)
Policías Comunitarias (PCs)
Revolution in Military Affairs (RMA)
transnational corporations (TNCs)
Unión Sindical Obrera de la Industria del Petroleo (USO)
United Nations Program on Development (UNDP)

1

Introduction

[F]or matters of war are, of all others, most subject to continual mutations.[1]

Miguel Cervantes, Don Quixote

The Revolution in Military Affairs (RMA) entails a sweeping redefinition of the field of strategic studies, one that rhymes with sea changes experienced in a wide domain of other realms. The RMA can be viewed as a synergetic bundle of concepts that is marked by technological change, social reorganization, and sometimes by a new system of thought, among other features. It is this synergy between various components of the RMA that has underpinned the strategic victories and losses for assorted social forces in Colombia and Mexico, and that best explains the transformation of security themes in both countries. While the RMA is usually discussed in relation to Middle Eastern politics or the military adventures of "great powers," this book veers off the beaten track to examine the manifestation of the RMA in two of the most strategically important countries in Latin America.

Mexico hosts a number of newfangled security themes. Some of these contributed to the development of quintessential features associated with the RMA, such as the elements of information warfare and global network organization that were harnessed by the postmodern Zapatista rebels. Further, we shall see that the highly successful reclamation of local strategic space by Guerrero's Policías Comunitarias (Community Police) exemplifies a truly grassroots approach to the RMA. There are also some cutting-edge dimensions to traditional themes such as migration. This is especially the case in the context of an increasingly militarized frontier, which has appeared in the wake of Washington's Fortress America response to 9/11. Especially interesting in Mexico's case are the contradictory influences entailed in the integrative effects under the North American Free Trade Agreement (Nafta) versus the isolation implied by a fortified border featuring longer walls, drones, and thousands of members of the U.S. National Guard.

For its part, Colombia has endured the most potent insurgencies in the Americas since the fall of Peru's Sendero Luminoso in 1992. Between the years 2000 and 2006, it has ranged between the third and the seventh largest recipient in the world of U.S. military assistance under Plan Colombia (PC). This strategic package has transformed the country into a hemispheric theater for the RMA, and as a bastion of military strength designed to suit American interests. It is remarkable that one finds in Colombia many of the same strategic themes one would expect to observe in parts of the Middle East, such as terror, asymmetric and privatized warfare, ultra-surveillance, and historically incessant violence with no clear resolution in sight.

Beyond the strategic significance of each country in its own right, Colombia and Mexico make good comparative case studies for an analysis of the relation between social forces and the RMA. They share commonalities that invite strategic generalizations, but are also marked by sufficient distinctions to evoke a broad range of analytical considerations. Regarding similarities, they both host top-tier security issues in the Americas and display essential elements of the RMA, as we shall see. Narcotrafficking, for example, has blurred the distinction between crime and war in both countries. They are among the very few Latin American states to host significant insurgencies that have spilled into the first decade of the new millennium, although the nature and context of these rebellions are highly distinct. Colombia and Mexico also have been Washington's staunchest strategic allies in the region, and are arguably the two most strategically significant countries for the United States in Latin America.

Yet they bear important distinctions with regard to political culture, geopolitics, stability levels, state formation and development, and so on. One prominent distinction between the two concerns the insurgents they have hosted. In Colombia, guerrilla movements and paramilitary forces have relied heavily on violence and even indiscriminate carnage. Colombian insurgents, at various points, have affected everyday life in most parts of the country. By contrast, Mexican guerrillas have been confined largely to pockets in the southern part of the country, and range from the nonviolent Ejército Zapatista de Liberación Nacional (EZLN) to guerrillas such as the Ejército Popular Revolucionario (EPR) and its offshoots, which have utilized carefully selected targets for relatively infrequent acts of violence. In Mexico, the only guerrillas that have had even a fleeting impact on what could be called national life are the EZLN, and this has been based on ideational power rather than on the Clausewitzian military might wielded by the Colombian guerrillas and paramilitaries. These differences between insurgents in Colombia and Mexico are related to broader distinctions between the generally stable, centered, and highly institutionalized politics of Mexico versus the chaotic, dispersed, and violent political economy of Colombia.

This book is as much about the RMA as it is about the case studies of Colombia and Mexico. That is, while there is a special interest here with regard to social forces in these two countries, there is also a focus on the phenomenon of the RMA itself and how its unfolding manifestations play out in distinct contexts. We shall see that although Colombia and Mexico host identical components of the RMA, these elements are filtered through the unique circumstances of each case. While the cases of Mexico and Colombia bear enough rough similarities to evoke comparison, it is particularly important for this study that they are sufficiently distinct to provide two poles from which to consider a wide range of implications linked to the RMA. We shall see that the Colombian case exemplifies an RMA from above, one implemented by the state, while the Mexican case in many ways demonstrates an RMA from below, one inspired by civil society. Further, Mexico demonstrates the manifestation of the RMA in a relatively stable country, while Colombia exemplifies its expression in the context of a weak state. Together, then, these cases illustrate a wide spectrum of themes related to the expression of the RMA in Latin America.

For whom does the RMA represent a revolution? While the state and belligerent forces will be examined in some detail, a prime consideration here is the often overlooked elements of civil society and the majority population of the poor. This endeavor will analyze key social forces in relation to the RMA, with respect to their role in helping to define the RMA as well as their predicament of being the object of its effects. A central argument of this book is that civil society can empower itself through a clear understanding of the RMA, and can help shape the emerging revolution rather than being victimized by it at the hands of adversarial forces. There will be an emphasis, then, on the dimensions of human security associated with this rupture.

In the Colombian case, there will be an analysis of the RMA in relation to the country's cast of potent belligerents, including the Fuerzas Armadas Revolucionarias de Colombia (FARC), the Ejército de Liberación Nacional (ELN), and the Autodefensas Unidas de Colombia (AUC). There will be a special focus on the plight of two elements of civil society on the front lines of the RMA in the country—the displaced population, which is the world's third largest, and the labor movement in Colombia, where it is the most perilous place on the planet to be a union member. Thus, there will be an emphasis on the role of civil society in relation to refashioned strategic struggle, in addition to the crucial and conventional examination of the state versus belligerent forces.

In the case of Mexico, there will be a special focus on the country's most "Colombianized" state, Guerrero. We shall see that Guerrero is a microcosm of sorts for many of the components of the RMA that are apparent in Mexico. In terms of belligerents there will be a focus on the EPR, a guerrilla group that has its roots in the notorious Dirty War era of the 1970s. Frequent references will be made to the Zapatistas in Chiapas and to a rising tide of powerful social movements in Oaxaca. With regard to civil society, the spotlight will be upon human rights nongovernmental organizations (NGOs) that have largely carried the torch in the struggle to promote human security, that is, a conception of security designed for grassroots forces in the context of repression. Another important manifestation of grassroots security is the local reclamation of strategic space achieved by the Policías Comunitarias. Although the magnitude of the strategic and social problems in Guerrero rivals and sometimes even surpasses those in Chiapas, it has not received nearly the global attention that Chiapas has and at times has appeared to be all but forgotten. We shall address the question of why social forces in Guerrero, and also in Colombia, have garnered relatively limited global attention, in contrast to Chiapas, where locals have proven to be experts in attracting the global gaze and at harnessing elements of the RMA in their favor.

The Context of the RMAs

Before we consider the current RMA and the related case studies of Colombia and Mexico, let us briefly visit the ghosts of other RMAs.[2] It should be noted at the outset that many of these rested on the appearance of a primary but sweeping change, often a technological one, which further spawned transformations in the conduct of warfare. One example is the Infantry Revolution, which occurred during the Hundred Years' War (1337–1453) and witnessed the displacement of the cavalry by the infantry. Perhaps its most important causal factor was the development of the longbow arrow, which rendered the cavalry vulnerable through its ability to pierce armor. This was followed by the Artillery Revolution of the fourteenth and early fifteenth centuries, which also featured a relatively narrow range of causes. This rupture was sparked by better technology in terms of longer gun barrels, by the greater economic efficiency of metallurgy, which reduced the cost of iron cannons by about one-third, and by cheaper and better gunpowder. The next jump was the Revolution of Sail and Shot, which, as the name suggests, was limited to developments concerning war on the seas. As one expert observed, "By 1650 the warship had been transformed from a floating garrison of soldiers to an artillery platform."[3]

Perhaps closest conceptually to the RMA we are currently witnessing is the one that began with the Napoleonic Wars, the remnants of which stretched into the twentieth century. Particularly noteworthy is the breadth of radical change that transpired during this epoch. The development of industrial capitalism and the birth of the modern nation-state, complete with patriotism and a national army, marked truly fundamental shifts in social organization and political space. Broad technological and industrial developments affected weaponry, exemplified by the reduction in the weight of cannons by 50 percent. Some observers see an analytical parallel between the proliferation in France of the printing press—such that periodical publications skyrocketed from 42 in May 1789 to 250 by the end of that year—and the appearance during the current RMA of the Internet and its role in the vast dissemination of knowledge.[4] Overall, crucial shifts in knowledge structures associated with modernity were fully apparent by the late eighteenth century—in terms of positivism and binary thought, as well as notions of progress, human nature, linear development, and so on. These underpinned the RMA of the era.

There are a number of milestones associated with the current RMA. The concept's fuzzy outlines first appeared in the Soviet Union during the 1960s and 1970s, with Moscow's preoccupation regarding the consequences of the development of precision nuclear weapons. Ironically, the RMA did not take clear shape until the fall of the Soviet Union and the dimming of the global contest between capitalism and socialism. Shortly afterward, the 1991 Gulf War represented an RMA watershed due to its reliance on satellite reconnaissance, near-real-time surveillance, and lethal precision weapons. In the Americas, Mexico's post-1994 Zapatista struggle featured the advent of information warfare with a twist, as it was mastered by a small, indigenous rebel group rather than by a superpower. And in Colombia, backward guerrilla groups transformed into mega-military machines through lucrative adventures in transnational crime. Despite all that and more, doubts about the existence of an RMA lingered right up to the beginning of the new millennium. These all but dissipated, however, with al-Qaeda's infamous attack on the U.S. homeland in 2001, and with the subsequent U.S. invasion of

Afghanistan, which made use of real-time satellite images that allowed the United States to direct the war from the MacDill Air Force Base in Tampa, Florida.

It was noted that the Napoleonic RMA is conceptually similar to the current RMA. This is because both cases exemplify a broad shift in strategic affairs that has moved in step with a sweeping social revolution. Just as the Napoleonic RMA appeared within the wider rubric of modernity—including the emergence of industrial capitalism, the modern nation-state, and newfangled philosophical principles such as linear progress, balance, objectivity, and binary thought—the current RMA rides the crest of postmodernity. It cannot be reduced to a single causal factor, but is the synergetic product of a number of key developments. That is, the context for the present RMA includes transnational capitalism and globalization, the dispersion of information and of weapons, the appearance of new political spaces (such as trade blocs, as well as autonomous regions based on ethnicity), the advent of identity politics that transcends traditional patriotism, ecocide, and epistemological considerations that celebrate multiple truths, nonlinear politics, network organization, and so on. Within the context of a broader social revolution, the current RMA signifies a shift both in the array of actors and in the type of conflict apparent in strategic affairs. The new RMA can be exploited by a vast array of players—including transnational criminal syndicates, NGOs, labor unions, newfangled insurgents, and autonomous communities, in addition to more traditional entities such as the state and international organizations. Further, the nature of the conflict has transformed to include a focus on phenomena such as asymmetric warfare, privatized war, epistemic fissures between combatants, and the blurring of war and crime. Let us explore these shifts in greater detail.

Conceptualizing Features of the RMA

While the RMA is about change, paradoxically it also entails a strong dose of continuity. That is, some classic strategic themes remain as important as ever, but need to be situated within a new context. For example, fundamental elements of strategy, such as time, space, and intelligence, among others, can be viewed through new prisms. So, too, can the timeless observations of classical thinkers such as Thucydides, who pointed to "three of the strongest motives, fear, honor and interest."[5] The remainder of this chapter will provide a brief conceptual outline of the key features of the current RMA. These include epistemological considerations, asymmetry, complexity, elements of political economy, ultra-surveillance, network organization, identity politics, as well as fear and terror. The analysis of these features will rely on insights from classic strategic texts in order to help discern what is old and what is new.

Rupture and Epistemology

The concept of the RMA necessarily has as its centerpiece the notion of rupture. On the one hand, it is important not to exaggerate the extent of change or to minimize the elements of continuity. But, on the other hand, it would be highly perilous not to appreciate the strategic implications of the winds of change when they actually do appear. We shall situate the RMA within a broader fabric of epistemological change, or what some deem to be a "system of systems" approach.

The concept of discontinuity is featured in various perspectives to global politics. It is prominent in classical realist and Marxist approaches, but most recently has been associated with postmodernism. This is especially so in light of Michel Foucault's seminal theoretical work *The Order of Things,* in which he examines "change in the fundamental arrangements of knowledge."[6] Despite the huge gap between postmodernism and Marxism, one quickly notices that Foucault points to the same ruptures as does Karl Marx.[7] A prominent difference is that Foucault emphasizes the epistemological elements of the rupture, or what he calls "episteme," while Marx emphasizes shifts in relations of production and class within his historical materialist framework. Insights from both approaches, emphasizing political economy and epistemology, can be helpful when analyzing the current RMA.

Long before Foucault and Marx, classical realists celebrated the notion of rupture in relation to epistemology. Thucydides, some 2,400 years ago, sought to tell the story of war and politics in vast distinction from his predecessors such as Homer. In Homer's work, analogy represents an essential epistemological component of strategic analysis, with human behavior resembling natural phenomena such as the formation of birds in flight or the patterns of the wind.[8] The gods also were perceived to play an important role in warfare, and so divinity is a crucial element of the story's episteme. For example, in the *Iliad,* combatants routinely prayed to God during battle,[9] and divine messages concerning warfare were transmitted through dreams.[10] But the gods were not always in full control of human behavior. At one point, Zeus went on a vacation of sorts to party with other gods in Ethiopia, leaving humans to settle their own affairs![11] While the gods may not have spent all their time at the steering wheel in Homer's tales of warfare, what is clear is that they were central to his story and to epistemological conceptions of political truth.

While Homer highlighted analogy, divinity, subjectivity, and how it felt to be engaged in armed struggle, Thucydides told the story of the Peloponnesian War by emphasizing temporal sequence, spatial location, and what he viewed as objective fact. Indeed, the opening sentence of the *History of the Peloponnesian War* stresses the importance he assigned to objectivity, as Thucydides places himself in the third person: "Thucydides, an Athenian, wrote the history of the war between the Peloponnesians and the Athenians."[12] Thus, accounts of strategic rupture have been apparent for at least 2,400 years.

Overall, the RMA is one facet of a systemic change that entails an epistemological jump. The present epistemic shift is reflected, for example, in the appearance of social forces motivated by new social "truths," in novel conceptions of political time and space, in conflicts linked to different systems of knowledge, in emerging types of social organization such as the cell and the network, in unfolding manifestations of political identity, in newfangled conceptions of enemy and threat, and so on. The strategic realm or "system" should be considered in the context of dynamic interactions among a network of other social systems—economic, cultural, organizational, political, technological, scientific, and so on.[13] The RMA, then, represents the strategic implications of a broad new episteme, or what some in the strategic studies literature have deemed to be a "system of systems."[14] Not only has the realm of strategic affairs witnessed a discontinuous jump, but it has done so in relative harmony with other fields. From this perspective, "forms of war" and "forms of life" are underpinned by the same system of thought.[15]

Asymmetry

Although history is peppered with David and Goliath stories, there is no question that a greater degree of asymmetric warfare represents one factor that helps define the present era. The two world wars of the twentieth century, as well as the cold war that followed them, featured a strategy among major powers that focused on enemies of relatively equal standing. For example, during the cold war, the United States developed weapons and shaped its armed forces to meet a perceived threat emanating principally from the Soviets. There was a great emphasis on air power, long-range precision weapons of mass destruction, and so on. Once the Soviet Union collapsed, Washington began to focus on China as a potential rival, and continued apace with its previously conceived notions of great power warfare.[16]

While warfare between nation-states of relatively equal power remains a distinct possibility, asymmetric combat has been on the ascendant since the 1990s. Terrorist attacks authored by al-Qaeda and others represent a classic example of this, but there are other important cases of asymmetry that are apparent in the Americas. Mexico's Zapatistas demonstrated that a poorly armed guerrilla group can get its message across and avoid defeat through an information war supported by a network of global NGOs. Colombia's assortment of guerrilla groups showed that transnational crime can provide generous support for a plethora of winning tactics associated with asymmetric warfare. Asymmetry implies differences between the types of contestants, and does not necessarily refer to power differentials. Further, we shall see that asymmetry suggests distinctions based on organization, weapons, episteme, and so on. While asymmetric struggle generally refers to the state versus nonstate belligerents, it is also helpful to consider asymmetry in relation to struggles faced by components of civil society—especially in cases where there exists a blurred distinction between war and politics.

Complexity

This element of the RMA suggests multicausality, as well as plural and sometimes unpredictable effects of given actions. Some of this may involve a fresh awareness of arrangements that were already complex but were not appreciated as such. Ecocide is a good example. It is now understood that the modernist notion of "man's" ability to conquer Mother Nature was unfounded. It has given way to the present view of a complex ecology whereby intense interaction among multiple agents yields unanticipated consequences that may not be immediately obvious. We shall consider this component of complexity with regard to the fumigation of coca crops in Colombia as well as the proposed construction of the La Parota Dam in Mexico.

Beyond ecological aspects, there are countless examples of complexity that are strategically relevant. Complexity, for example, is also spawned by revolutionary changes in communication, prompting a bombardment of information and surveillance techniques with consequences that are still unfolding. This ultra-dissemination of information is compounded by the dispersion of weapons. The ideal of a Leviathan's monopoly on force has given way to a reality whereby, in Colombia's case, the local population has more weapons than the state, guerrilla, and paramilitary forces combined. This perverse predicament of the decentralization of arms has had the ironic effect of "democratizing" violence while democratic political structures take a beating.[17] Global connectivity in the financial realm also raises complications, as demonstrated by the widespread "tequila effect" of Mexico's 1994–1995 economic meltdown and by Colombia's stock market roller-coaster ride of mid-2006. Further, complexity is exemplified by the spin-off effects of extremely lucrative crime, such as narcotrafficking's corruption of government officials at a juncture when the state is already too anemic to perform its duties properly.

Overall, there are numerous elements of complexity associated with the current RMA.[18] Catalysts of this phenomenon include economic globalization, an overwhelming barrage of information, the presence of epistemological gaps, the dispersion of force, the redrawing of political space, the formation of new social identities, a trend toward asymmetric warfare, the blending of crime with war, and others. While life has never been simple, perhaps the strategic web is becoming even more complex.

The Political Economy of the RMA

> Power is tolerable only on condition that it mask a substantial part of itself.[19]
>
> *Michel Foucault,* History of Sexuality

Inequity

It takes more than class conflict and inequity to promote insurgency. After all, there are many venues in the world where these factors exist, but where guerrilla groups and the like are absent. But it is important to recognize that revolutionary movements of the Latin American sort would be nonexistent in the absence of extreme socioeconomic inequity—in fact, the region has the most unequal distribution of wealth in the world.[20] Vast inequity, which generally has worsened in Latin America since the debt crisis of the 1980s, is one crucial factor among others that promotes insurgencies. This broad theme is not limited to leftist political economists. It is emphasized in much of the classical realist literature.

Thucydides, for example, emphasized the politics of a vertical tension between exploiters and the subjugated. This tension meant that the "haves" needed to remain constantly on guard for inevitable attempts by the subjugated to reverse the inequitable order of things. Related to this, he also referred to what postmodernists would later term the "inside/outside duality,"[21] as within its borders Athens "favor[ed] the many instead of the few . . . (with) class considerations not being allowed to interfere with merit."[22] But outside of its borders, Athens pursued a strategy of political and economic domination: "For the love of gain would reconcile the weaker to the dominion of the stronger, and the possession of capital enabled the more powerful to reduce the smaller cities to subjection."[23] Under such circumstances resistance is predictable, both at the outset in an attempt to quash the conquest, and afterward, if imperialists initially succeed.[24]

Thucydides was not alone in his recognition of the link between strategic worries and highly inequitable economic situations. Sun Tzu II warned that rulers were likely to be deposed in such situations,[25] as did Machiavelli when he underscored the perils of regimes that benefit the few and injure many.[26] For other realist writers, the strategic implications of inequity are more implicit than explicit. Hobbes, for example, celebrates the element of consent when he emphasizes "the mutuall transferring of Right" that underpins the social "contract" and hence the Leviathan.[27] He is famous for his assertion that "men have no pleasure, (but on the contrary a great deale of griefe) in keeping company, where there is no power to over-awe them all."[28] Part of what characterizes political development is precisely the existence of a Leviathan that is underpinned by a consensual social contract. Conversely, the profound socioeconomic inequity that exists in regions such as Latin America can preclude consent, and therefore prohibit the development of a social contract as well as a state monopoly on the use of force in the form of a Leviathan. That is, why should an impoverished majority population consent to a strategic contract that protects and perpetuates the extreme wealth and privileges of a tiny minority? Under such circumstances, class consciousness and polarization are likely to be prevalent, contributing to the likelihood of armed insurgencies.

Privatized Warfare: Neomercenaries and Contractors

The increasing prevalence of privatized warfare is a reflection of wider economic and social forces. Since the 1980s there has been a growing trend toward the privatization of many components of the social sector. These include privatized education and health care, privatization of cooperative or communal land and of government corporations, the practice of "contracting out" work from government institutions, and so on. In much of Latin America, such changes may represent prescriptions contained in the International Monetary Fund (IMF) debt restructuring packages. Through this model, smaller government and less state intervention have been viewed as key goals. Much of this shift has reflected the growing power and interests of transnational capital, often under the banner of "globalization."

One crucial aspect of the privatization of war is the proliferation of "neomercenaries" or "private contractors," and this is particularly important in the Colombian case. The term "mercenary" is fraught with connotation. In terms of ordinary usage, mercenaries are usually described as private soldiers who freelance "their labour and skills to a party in a foreign conflict."[29] They may also be described as "individuals or organizations who sell their military skills outside their country of origin and as an entrepreneur rather than as a member of a recognized national military force."[30] Falling under that definition, for example, are large corporations such as Military Professional Resources Incorporated (MPRI), located just outside of Washington, DC. MPRI provides a wide array of services, such as tactical, operational, and strategic advice for the structuring, training, equipping, and employment of armed forces. It also provides services related to strategic planning, force development, research, intelligence, and electronic warfare.[31] Another variant is more specialized corporations, for example, ones that specialize in surveillance or air-based activities.[32] Beyond these variations is one crucial distinction regarding whether the corporation provides offensive or defensive services. "Private military companies" engage in offensive operations, and "private security companies" provide defensive services.

Despite the wide variety of entities that come under the broad banner of mercenaries, the legal definition of the term is very narrow and exclusive. Often the correct legal terminology for what might otherwise be called neomercenaries is "contractors." In terms of international law, Article 47 of the Geneva Protocol (1949) remains the definitive text. It defines a mercenary as someone who

> (a) is specifically recruited locally or abroad in order to fight in an armed conflict; (b) does, in fact, take direct part in the hostilities; (c) is motivated to take part in the hostilities essentially by the desire for private gain and, in fact, is promised, by or on behalf of a party to the conflict, material compensation substantially in excess of that promised or paid to combatants of similar ranks and functions in the armed forces of that party; (d) is neither a national of a party to the conflict nor a resident of territory controlled by a party to the conflict; (e) is not a member of the armed forces of a party to the conflict; and (f) has not been sent by a state which is not a party to the conflict on official duty as a member of the armed forces.[33]

This definition offers a number of loopholes for those wishing not to be associated with the taboo term of mercenary. Rather surprisingly, it is item "c," which specifies that the mercenary must be motivated by private gain, that often represents the biggest legal roadblock to using the term. This is because it is difficult to prove in court that profit is truly the motivation. Further, item "b," which notes that the participant must take "direct" part in hostilities, has been interpreted strictly as meaning offensive action in combat. It would not include, however, defensive private soldiers, military trainers, intelligence services, and so on.

Mercenaries, or private combatants, are not new to the strategic landscape. As one observer notes, "For at least three thousand years mercenarism has been a feature, often the major feature, of the institutions of organized violence."[34] In 334 B.C., for example, Persia utilized Greek mercenaries in the war against Alexander the Great. The use of private warriors waxed and waned over the centuries. England attempted to counter Spanish naval superiority in the Americas with private military operations led by the Sea Dogs, some of whom included Francis Drake and Walter Raleigh.[35] Elsewhere, the "Condotieri," or military contractors, were quite common in Italy by the middle of the fourteenth century, partly due to a surplus of soldiers in the context of truces in the Hundred Years' War.[36] This remained a dominant trend there and elsewhere in Europe through the 1500s. The situation sparked Machiavelli's sharp criticisms of mercenaries in chapters 12 and 13 of *The Prince*, in which he refers to them as pernicious and "useless."[37] But Machiavelli was ahead of his time. Mercenaries did not fall out of fashion until well into the modern epoch. The development of modern nation-states, especially after the American and French revolutions, meant the construction of a Leviathan complete with a national army motivated by patriotism.

But mercenaries never fully disappeared. The French Foreign Legion, which one observer called the "most famous of all mercenary bands," was active in Africa especially after 1830. England hired mercenaries from Nepal and elsewhere during various imperial adventures in the nineteenth century.[38] In the United States, contractors outfitted British forces during the War of Independence, and these continued to be used after the Revolution by a host of American leaders, including George Washington.[39] In the American Civil War, contractors were used extensively by Confederate forces.[40] They continued to be utilized indirectly by the United States throughout the twentieth century, during World War II, for example, and also during the Vietnam War, where there were 52,000 non-Vietnamese contractors working with U.S. interests.[41]

Throughout the cold war, there was an assortment of anti-Communist mercenaries working in developing countries. This was especially the case in parts of Africa where the processes of postcolonial national development and the formation of national armies were the weakest. In what would become a commonality, mercenaries worked to protect the valuable extractive sector in Africa. This general phenomenon commenced in the 1950s on behalf of the diamond company De Beers, and continues apace into the twenty-first century.[42] In fact, mercenaries and private security companies thrive off the extractive sector in developing countries—in items ranging from diamonds and emeralds to oil and wood.[43]

The pace of neomercenary activity, or of the use of contract workers, accelerated sharply during the 1990s, beginning with the Gulf War. Indeed, the United States reversed the reputation it earned in 1985 as being "almost the only imperial power in history that has never employed mercenary troops directly" to its status in 1999 as being "the major contractor of mercenary services in the world."[44] In Operation Desert Storm, there was one contractor for every 100 military personnel, while a few years later, in Bosnia, the tempo increased to an astonishing ratio of 1 to 1. Between 1994 and 2004, the Pentagon signed contracts worth about $300 billion with 12 companies for the services of military contractors. Part of this included the provision of private security in Colombia, as we shall see. The U.S. invasion of Iraq in 2003 prompted Washington's largest ever reliance on private security apparatus. Private contractors, for example, have provided security for the chief of the Coalition Provisional Authority L. Paul Bremer III, as well as for other top officials. They have also escorted supply convoys through hostile territory and have defended key U.S. installations.

This phenomenal growth since the 1990s of contract workers, or neomercenaries, is due to three broad factors. The first concerns a drop in the U.S. armed forces, precipitated by budget cuts following the fall of the Soviets. Between 1989 and 2001, the U.S. government cut uniformed military personnel by 38 percent, and the Department of Defense trimmed civilian employees by 44 percent.[45] The second factor has been an ideological preference for privatization that rhymes with other features of globalization. Finally, as in the Colombian case, the use of contractors is meant to attract less public attention to U.S. policy than if American military forces were used instead.

Paramilitaries, Crime, and the Privatization of Warfare

Paramilitary forces have been present in Latin America for centuries. In the recent past, paramilitaries were a noticeable feature of Latin American warfare throughout the cold war.[46] Noteworthy in this regard was the creation by the Reagan administration of the Nicaraguan "Contras" in the 1980s. The Contras were a paramilitary force developed at an initial cost of $19.5 million with the objective of bringing down the socialist Sandinista government and were specifically designed for the purpose at hand. Not only had the international community branded the U.S. intervention in Nicaragua as illegal,[47] but also part of the funding for the Contras was mired in crime. For example, the Kerry Commission discovered that Colombian cocaine was transported to the Costa Rican ranch of a U.S. Central Intelligence Agency (CIA) agent, then sold in the United States, with proceeds financing between $50 and $75 million for the purchase of arms for the Contras.[48] The elements of crime and paramilitary activity also pertain to the Colombian and Mexican cases, as we shall see.

Paramilitaries come in a number of stripes, but what is common among them is that they are nonstate military forces operating with a right-wing political agenda. They are privately funded and serve private interests. In Mexico, these have ranged from small, shadowy groups that represent the illicit security arm of local economic elites and political figures, to highly organized and publicized units with a clear political agenda. In Colombia, they have been an extremely disciplined political militia comprising tens of thousands of soldiers and funded chiefly through narcotrafficking. By 2007, a serious scandal emerged around mounting evidence confirming what many had long suspected—that the *paras* were closely tied to key elements of the Colombian government and to U.S.-based corporations operating in the country. Simultaneously, the paramilitaries were deeply engaged in a "demobilization process" that at times has appeared to be shell game, a point that is developed in Chapter 4. In general, the right-wing political interests of the paramilitaries often have rhymed with the agenda of the state or of global actors such as the United States. Because the paramilitaries officially operate separately from the state, government officials can claim they have no control over such forces—thereby distancing themselves from *paras* who typically commit human rights abuses while playing dirty to achieve shared interests with the state. Related to this, *paras* often act with impunity. It is worth emphasizing that the United States and its Latin American allies have not attempted to curtail paramilitary groups with nearly the vigor they have demonstrated vis-à-vis leftist guerrillas.

Crime, especially narcotrafficking, is a hugely important dimension of the privatization of warfare. The development of this industry has accelerated enormously since the debt crisis and the imposition of neoliberal structures in the 1980s. As jobs disappeared, as banks dove into crisis, and as welfare programs were slashed, narcotrafficking filled gaps. All this was driven by Northern consumption, a point to which we shall return. Illicit drug trafficking and other types of crime have corrupted government officials as well as the police and military. That is, it has severely weakened legitimate security structures. Not only has it provided the primary source of funding for right-wing paramilitaries such as Colombia's AUC, but it has also financed the activities of leftist guerrillas. These include the hemisphere's most potent guerrilla groups, including Colombia's FARC and, to a lesser extent, the ELN. With regard to Guerrero, Mexico's secretary for Public Security and Citizen Protection in May 2006 deemed narco-related violence in Acapulco and elsewhere to be far worse than the violence caused by guerrillas in the state during the 1990s.[49] Overall, narcotrafficking and other crimes have handsomely financed some of the most important subversive groups in the hemisphere. The blurred border between war and crime is a significant dimension of the RMA and of the privatization of warfare.

Consumption and the Privatization of Security

While much attention rightly has been afforded to matters surrounding production in relation to social power, one should not ignore the crucial theme of consumption. Northern consumption patterns often determine much of what is produced in the South.[50] Northern consumption is related to the trend toward the privatization of security as well as to more generalized strategic problems in the two cases considered here. In terms of legal goods and services, oil, minerals, and jewels top the list. With regard to the energy sector, oil pipelines in Colombia, especially in chaotic and war-torn Arauca state, have been the site of fierce contestation among assorted private interests. These include private military and security corporations hired by transnational oil companies, as well as paramilitaries and guerrillas who depend on extorting the industry to finance their subversive activity. The oil flowing from the renowned Oxy pipeline is largely destined for export to service Northern energy needs. In Mexico, the production of electricity for transnational corporate interests proposed under Plan Puebla Panama (PPP) has been bitterly opposed by guerrilla groups such as the Zapatistas and the EPR, as well as by a variety of social movements. If this project were to reach fruition, security problems would likely develop.

Northern markets for minerals, precious jewels, and illicit drugs have also underpinned privatized warfare and a host of security problems. This is particularly the case in Colombia, where extortion from various mines has been a key source of funding for guerrilla groups and paramilitaries. It is also worth noting that some of the key figures who have dominated Colombia's narcotrafficking industry have dealt in the contraband emerald market in the past. Turning to illicit narcotics, it is precisely their illegality that bestows on the industry its astonishing wealth and power. In the context of prohibition, "power and desire are joined to one another."[51] Here the allure of the prohibited is enhanced, and the power relations that stream from the illicit product are laced with privatized violence. It is important to underscore that the industry is almost completely driven by Northern consumption and demand. Thus, Northern "cokeheads" unwittingly make regular donations to the carnage produced by Latin American paramilitaries, guerrilla groups, and transnational crime syndicates. Overall, patterns of consumption are key to strategic affairs in both Colombia and Mexico.

Intelligence and Surveillance: Roots, Utility, and Limits

The RMA has occurred in tandem with a revolution in information systems, especially with regard to instant satellite surveillance and Internet communication. This has been apparent since at least the early 1990s, with the use of near-real-time satellite surveillance in the Gulf War, and of infowar and netwar by the Zapatista rebels during their post-1994 struggle.[52] Technological breakthroughs have precipitated the availability of a far greater range and volume of information, and the ramifications of these breakthroughs are still unfolding.[53]

Within the current RMA, the power of surveillance is compounded because, for its ultimate expression, it must be continuous. In fact, this element distinguishes the narrower facet of reconnaissance from the broader (and continuous) manifestation of surveillance.[54] The importance of the feature of continuity was grasped more than two centuries ago by Jeremy Bentham in his discussion of the Panopticon. He observed that "the more constantly the persons to be inspected are under the eyes of the persons who should inspect them, the more perfectly will the purpose of the establishment have been attained . . . the most important point (is) that the persons to be inspected always feel themselves as if under inspection."[55] This notion was developed further by Foucault, who focused on the power of continuous surveillance in the disciplinary sense, that is, for the control by institutional authorities over individuals or groups.[56] But, due to the asymmetric nature of the RMA, surveillance also can be used in the name of liberation by small groups over large nation-states. At any rate, the continuous nature of surveillance expands even further the field of information to be interpreted and digested.[57] What kind of infrastructure and level of person-power are required to cope with the provision of constant information? The continuity of surveillance represents both a strategic boon and a daunting challenge.

While today's system of surveillance may rely heavily on high-tech gadgets, one should not underestimate the staying power of human intelligence[58]—which includes the strategic use of specialized spies or the harnessing of local and regional eyes. This classic security theme has been celebrated throughout the ages. For example, in the *Iliad,* Homer expounded upon the importance of spies and surveillance in the context of war.[59] So, too, did Sun Tzu, who insisted that "great success" depended on good intelligence, a topic to which he devoted an entire chapter of *The Art of War.*[60] He suggested that "[f]oreknowledge cannot be gotten from ghosts and spirits, cannot be had by analogy, cannot be found out by calculation. It must be obtained from people, people who know the conditions of the enemy."[61] Further, Miyamoto Musashi, in the Japanese strategic classic *The Book of Five Rings,* underscored the importance of intelligence and surveillance when he observed, "To see the faraway as nearby . . . is essential to the martial arts" and to warfare.[62] Through intelligence, one can determine the context of the struggle, the likely strategy and tactics to be pursued by adversaries, and so on.

With regard to Colombia and Mexico, we can outline two general purposes behind the use of surveillance/information in relation to the RMA. The first concerns its use in a strategic-military sense. The governments of Colombia and Mexico, with the support of the United States, have employed a variety of innovative and elaborate surveillance systems to weaken the power of subversive groups. Many of these efforts have met with success. But the use of surveillance/information is a two-way street. First, social movements, NGOs, and insurgents have relied heavily upon it, with varying results, as we shall see throughout the book. Second, the crucial strategic use of surveillance/information goes far beyond the strictly military realm, and is important in areas such as political economy. Constant surveillance and provision of information affords Northern financiers and speculators important disciplinary power over Mexico, Colombia, and other Southern countries. If Northern investors or financial experts do not like what they see, for example, they can push the "sell" button on their computers, sending currencies, stock markets, and debt ratings of developing countries into a rapid dive. While the instruments of surveillance/information can be used in either a strictly military or economic sense, the practical implication of their usage is that they operate in tandem with synergetic results, as we shall see.

Despite the shine of surveillance satellites and the Internet as new tools of power, their limitations require full appreciation. Some of these relate to the theme of intelligence in general. In an historical sense, there have been some formidable detractors of the strategic use of intelligence and observation. Shortly before Clausewitz complained that weak intelligence sources represent a contributing factor to the fog of warfare,[63] Napoleon observed that "[i]n war, spies, intelligence count for nothing and it would be to risk the lives of men on weak calculations indeed to trust in them."[64] Despite the increasing accuracy of intelligence and surveillance that characterizes the current RMA, and the vast amounts of information available through high-tech surveillance, Napoleon had a point. One has to be sure to gather the appropriate information, situate it within the proper historical and epistemological context, and view with a healthy dose of suspicion the information provided by human intelligence.[65] When this does not occur, devastating failures result. For example, intelligence experts point to a long list of crucial U.S. intelligence mishaps over the years. These include the failure to anticipate a number of events: the fall of the Soviet Union, the appearance of the Zapatista guerrilla group in Mexico in 1994, the March 1995 sarin gas attack by the Aum Shinrikyo cult on the Tokyo subway system, India's May 1998 nuclear weapon test, North Korea's 1999 ballistic missile test, terrorist attacks on U.S. embassies and military personnel in Africa and Saudi Arabia in the 1990s, and, of course, 9/11.[66] Successful intelligence, then, depends on knowing which objects require attention. Further, within much of the relevant literature, there is still a propensity to emphasize the capacity of nation-states to use surveillance, therefore downplaying the important potential for asymmetric information warfare.[67]

It is worth emphasizing that there remain important objects that escape the human gaze, and information that is unavailable. As we shall see when we turn to the Colombian case, real-time satellite surveillance has some technological limitations. For example, the U.S. surveillance system has the capacity to detect certain objects and movements, such as caravans of moving vehicles, with an aerial resolution of one meter, even at night and during bad weather.[68] But it cannot detect the movement of people under such conditions. Further, surveillance devices cannot penetrate the flora and fauna of the Colombian jungle, where foliage can be dense and rise to many dozens of feet. Cloudy weather can also limit surveillance of human location and movement. Thus, venues that are typically cloudy and endowed with dense jungle foliage provide good protective cover for guerrilla groups and other subversive movements wishing to hide from high-tech surveillance devices.

Overall, it is ironic that classic strategic themes such as surveillance and intelligence represent a clear highlight of the present RMA.[69] In the Americas, the newfangled expression of these age-old themes first appeared with the United States' deployment of single-channel tactical satellites during the 1989 Operation Just Cause in Panama, and continued apace with its reliance on the U.S. Space Command during the occupation of Haiti in 1994.[70] From a grassroots perspective, the saga of the Zapatistas' revolutionary use of surveillance since 1994 raises key questions that will receive much attention in this endeavor: What attracts the Northern gaze, what does not, and what are the strategic implications of this?

Organization, Identity, and the RMA

There are a multitude of elements concerning organization that are relevant to today's RMA. Some of these are classic themes rewritten with a postmodern twist. Napoleon, for example, was keen on what today's RMA experts call swarming. He observed that "[t]he Art of War is to dispose one's troops in such a way that they are everywhere at once."[71] Like Napoleon, Jomini viewed strategy as an art, and suggested with regard to tactics and organization that "the principal thing to be attended to is the choice of the most suitable order of battle for the object in view."[72] Here the notion of "suitable order" suggests a wide menu of various organizational elements to deal with newfangled problems on a redefined battlefield.

Political culture and identity politics can have an important influence over appropriate strategic organization. Fuentes, for instance, analyzes the relation in Spain between a political culture of individualism and the historic development of a fresh military organization in the form of guerrilla warfare. In his brilliant analysis, Fuentes notes that the Spanish were

> [s]trong people, yes, but also extremely individualistic, as the Romans learned when they invaded the peninsula in 200 B.C. They quickly realized that the Iberian armies were brave indeed but ineffectual, because every man fought for himself and resisted integration into a larger unit and obedience to absent commanders or abstract rules. . . . Discovering that their strength lay in defense, the Spanish refused to offer a visible front line and instead invented guerrilla warfare. Surprise attacks by very small bands, preferably at nighttime; armies that became invisible by day, blending into the whitewashed villages and the gray mountainsides; dispersion, counterattack—these made up *la guerrilla*.[73]

More broadly, the point is that organization reflects political identity. In many parts of the world there has been an obvious shift toward the emergence of an identity politics that is quite distinct from loyalty to the nation-state. This implies the development of novel forms of political organization.

While the notion of organization represents a classic strategic theme, the question here concerns its specific manifestations within the current RMA. Social movements and insurgent groups increasingly have been organized into cells, networks, and other structures that defy traditional spatial and political boundaries.[74] This has been fueled by cyber-connectivity and by redefined political identity. With regard to belligerent groups, there has been a trend toward the dispersion of armed force, rather than the previous emphasis on centralization. The goals of rapid action and all-terrain mobility have been reflected in organizational structures associated with the RMA. A truly revolutionary change concerns the privatization of warfare and the development of related organizational shifts. We shall explore these and other organizational transformations in the chapters ahead.[75]

Fear, Terror, and the RMA

> Force can always crush force, given sufficient superiority in strength or skill. It cannot crush ideas.[76]
>
> *B. H. Liddell Hart,* Strategy

Classic realist literature has long considered the strategic implications of the use of force, as well as its relation to fear and terror. Thucydides, for example, analyzed multiple dimensions of force throughout the Peloponnesian War. In the context of the Melian debate, Thucydides recounted Athens' arrogant message to the Melians that "you know as well as we do that right, as the world goes, is only in question between equals in power, while the strong do what they can and the weak suffer what they must."[77] Although that battle was an important one, it is crucial to bear in mind that Athens' ultimate defeat in the war was partly the result of a backlash among its slave colonies and other weaker powers that resented Athens for maltreatment and therefore aligned with rival Sparta. There is a tense and fragile domination on the part of those who impose force, rather than the stable and enduring climate associated with consensual relationships.

Related to this are the moral implications regarding the use of force. In his provocative work *The Wretched of the Earth,* Franz Fanon argues that the colonized and exploited populations of the South are absolutely justified in using violence, because their subjugated position was imposed in the first place by violence on the part of Northern colonizers. He suggests that the colonized populations' "first encounter was marked by violence and their existence together—that is to say the exploitation of the native by the settler—was carried on by dint of a great array of bayonets and cannons."[78] Further, he argues that "it is the intuition of the colonized masses that their liberation must, and can only, be achieved by force."[79] For the oppressed, "violence is a cleansing force. It frees the native from his inferiority complex and from his despair and inaction; it makes him fearless and restores his self-respect."[80] Here we have a bottom-up view of the sort of vertical tension noted by Thucydides in the Melian debate—that is, an inverted rendition of the principle that "might is right."

Terror is the amplification of fear. One interesting dimension of this is Hobbes' observation that a "feare of things invisible" fostered the human creation of religion: "the Gods were at first created by human Feare."[81] On this point he was not alone. Even one of his fiercest critics, Giambattista Vico, suggested that humankind's first form of social terror was expressed through an intense fear of God due to inexplicable and relentless natural disasters within a state of nature.[82] This was the foundation of the first "system" of human organization and epistemology as noted by Vico.[83] He observed that, ironically, humans created gods due to their fear of the unknown (especially the appearance of disaster), and that this in turn led humans to have "a terrible fear of the gods whom they themselves constructed."[84] Extrapolating this to the current RMA, fear of the unknown is a key element of terror—that is, fear that induces intense paranoia from the prospect that terrorists could strike with devastating force at unanticipated times and places.

Napoleon begins his *How to Make War* with the observation that terror is a fundamental part of revolution and social rupture. He states:

> General rule: never a social revolution without terror. . . . How can one say to all those who fill up all the administrative organs, hold all the posts, enjoy all the riches; Get out! It is clear that they would defend themselves; it is thus necessary to strike them with terror, to put them in flight, and it is this which lamppost justice and popular executions have accomplished.[85]

Adding to this, Foucault suggests that "the ceremony of punishment, then, is an exercize of terror," as it is designed not only to force pain on the body of the specific victim(s), but also to serve as a spectacle to induce a sense of terror and paranoia among the larger target population.[86] Currently, the media is the primary instrument through which such terror is inflicted. That is, national and global terror would not exist to the extreme that it does if the media did not make it a spectacle through a barrage of images that are unavoidable for society at large.

Although terror is not new to the strategic landscape, there are a number of novel aspects associated with it that are apparent in the RMA. Terror has become an instrument of asymmetric war that affords nonstate groups far more military power than would otherwise be the case. It also permits such groups the capacity to strike Northern countries on their own turf, as al-Qaeda has demonstrated. Perhaps the effect of this development has been the most dramatic in the United States, a country with great experience in military adventures abroad but unaccustomed to coping with attacks to its homeland.

This irony has not escaped the notice of Latin Americans, who have witnessed in their own countries too many episodes of terror, on the part of both the state and insurgent groups. Terror has been employed by government forces in numerous cases[87]—some of these include Chile during the 1973 coup, Argentina and Mexico during the Dirty War of the 1970s, and certain Central American venues during the 1980s. With regard to insurgent forces, Peru's Sendero Luminoso ranked during the late 1980s and early 1990s as one of the world's most potent perpetrators of terror. Flashes of terror persist in the two countries under examination here. In terms of nonstate actors, Colombia has witnessed serious episodes of terror on the part of insurgents during the advent of the RMA, a phenomenon subsidized by transnational crime such as narcotrafficking. Mexico also has suffered from narco-terror.

Since 9/11, the word "terrorist" has become a code for Washington and its allies that is synonymous with utter political contempt, similar to the way the word "Communist" was used during the cold war. With regard to Latin American security, armed enemies of U.S. interests are increasingly being branded as terrorists, with little mention of ideology. Thus, Colombia's FARC and ELN, and Mexico's EPR, have been recast in government discourse as terrorist groups rather than as Communist insurgents. From the government's perspective, focusing on the alleged means of subversion (terror) rather than on the group's ideology is an instrument to discount entirely the group's platform. As we shall see with respect to Mexico and Colombia, nonstate actors as well as the state have participated in the deployment of terror in novel ways.

Organization of the Book

Overall, the RMA involves newfangled expressions of classic strategic themes, ones that are uniquely manifested through the filters of a particular country or region. In the subsequent analysis, three chapters will be devoted to each of the Colombian and Mexican cases. The first chapter in both of these sections will provide an historical overview from which to consider aspects of the present RMA. The remaining two chapters in each section will comprise an analysis of various components of the RMA, including epistemology, asymmetry, complexity, political economy, surveillance, identity, and organization, as well as fear and terror. It is the synergy among these various elements of the RMA that helps explain vast strategic changes in both Colombia and Mexico. The final chapter will be devoted to a comparative analytical conclusion.

2

Historical Aspects of Colombia's Strategic Landscape

The Colombian case is arguably the most important strategic crisis in the Americas at the dawn of the new millennium. It is Latin America's epicenter of brutal warfare, narcotrafficking, and population displacement. Along with the Israeli-Palestinian war, and the contest between India and Pakistan over Kashmir, the Colombian imbroglio is among the oldest military conflicts in the world. Yet, except for the odd "sound-bite" or perhaps a brief story buried in the back pages of the international press, it has remained off the media radar screen. And despite the eclipsing fears of global terror and the quagmire in Iraq, Colombia is among the largest recipients of U.S. military assistance globally. For example, it ranked seventh with regard to U.S. foreign aid in 2005[1] and was the third largest recipient of U.S. military aid for a number of years between 1998 and 2004. Colombia, then, deserves more scrutiny than it has received. We shall explore Colombia's strategic significance in relation to the RMA as well as the intriguing question of why this country has been relatively hidden from the global eye.

The spotlight here shall be upon critical social forces within Colombia. That is, in addition to the strategic implications of technological wizardry and the political interests of states, what does the RMA mean in terms of human security? There will be a special emphasis on labor unions, which historically have represented the most politically organized component of Colombian civil society, as well as on the war-ravaged displaced population, which is the world's third largest as of 2007. The country's three insurgent forces also will be examined.

This chapter will provide a brief historical portrait of the social struggle in Colombia up through the year 2000, beginning with an analysis of the two outstanding features, violence and fragmentation. We shall also consider key ruptures or RMAs over the course of the country's history. The rest of the chapter is devoted to an historical sketch of key social forces. This will provide the context from which to consider them in relation to the new millennium's Plan Colombia (PC) and the RMA which are the topics of the next two chapters.

Violence

There have been a plethora of excellent historical analyses of Colombia, which offer important empirical information as well as provocative conceptual frameworks.[2] Throughout these works, one notices a number of common themes. These include Colombia's extraordinary record of violence and the country's pronounced fragmentation, which together have produced numerous woeful effects. Given that Colombia's neighbors have been spared its level of violence, causal explanations must veer away from those that concentrate on the homogenizing notion of human nature and instead focus on the social structures that have bred such conditions. We shall begin with a brief chronicle of the violence that Colombians have endured, and then turn to an analysis of the reasons behind this in a subsequent discussion of fragmentation.

Colombia is the only country in the world to host the academic discipline of "Violentology." Although this is a relatively recent phenomenon, the record of violence in the country has deep roots. Historians, for example, have pointed to the virtually incessant warfare between the indigenous populations of Colombia during the preconquest period.[3] As was the case throughout Latin America, most of the indigenous population succumbed to disease once the Spanish arrived, rendering them relatively easy targets during the conquest. After the initial carnage associated with the Spanish invasion, colonial rule generally minimized violence in what is today Colombia. But there were some important exceptions. Perhaps chief among these was the bloodshed associated with the black slave trade, which began with the Spanish and continued into the midnineteenth century.[4] Following independence, almost constant civil war between the Liberals and Conservatives led to the death of 35,000 Colombians during 1820–1879, a figure that would equate proportionately to approximately 5 to 10 million deaths during the last 50 years of the twentieth century.[5]

The culmination of violent feuds and civil wars between the Liberals and Conservatives during much of the 1800s was the renowned War of 1000 days from 1899 to 1902, which marked the largest civil war in Latin America during the nineteenth century.[6] Somewhere between 80,000 and 200,000 Colombians lost their lives during that imbroglio. That exhausting ordeal appeared to deplete the propensity toward violence in the country until another horrific round of carnage slowly simmered and then exploded with La Violencia during the period 1948–1958.[7] At least 200,000 Colombians lost their lives in that final contest between the Liberals and Conservatives, which one astute observer has deemed as "one of the most intense and protracted cases of widespread collective violence in contemporary history."[8]

Beyond the extraordinary extent of violence unleashed during that episode, two features stand out that will prove significant for the subsequent analysis. First, keen observers of La Violencia emphasize the element of criminal economic enterprise as a significant factor, to the point that it conjured images of the Sicilian mafia for renowned historian Eric Hobsbawn.[9] Thus, the relation between crime and warfare in Colombia is far from new. Second, during La Violencia, large portions of the population identified with the combatants, in contrast to recent belligerents, who have commanded relatively little popular support nationally and have been perceived as victimizing the majority population of innocent Colombians.[10]

With La Violencia in the rear-view mirror, Colombia during the 1960s continued to witness the highest rate of violent deaths in the world. The country has remained near the top of that list into the new millennium. Although this remarkable level of violence has not altered significantly, there have been some important contextual shifts over the last few decades. One of these has been the acceleration of the war economy.[11] Propelled principally by the drug trade, and to a lesser extent by other crimes, continuous violence and warfare have been associated with the protection of immense economic bounties for belligerent forces, as we shall see. Perpetually high levels of violence, the criminalization of war, and the shifting spatialization of warfare, together with the lack of popular support for belligerents, have contributed to flashes of terror in Colombia since the 1980s.[12] While the abuse and overuse of the word "terror" and the discourse surrounding it will be discussed at length later in the book, the concept nevertheless represents an accurate depiction of the Colombian scenario.

Given this historical record, the themes of violence and war have found frequent expression in Colombian art. Fernando Botero, famed as perhaps the best painter in Latin America, captured these themes in a series of paintings he began in the 1990s and released in the new century. Moreover, the country's celebrated author, Álvaro Mutis, writes of the "nameless faces of hospitable, amiable people: all slaughtered by anonymous men . . . ready to take orders from those above them who work the puppet strings of implacable greed."[13] Beyond the place of carnage in Colombian literature, it is worth emphasizing that violence can serve the same function as language. It is meant not only as a message to its direct victims. Indeed, the victims embody a wider political statement that they are meant to transmit to society as a whole. Typically, the message is to remain silent, excluded, and isolated, and to stay off the radar screen of belligerent forces—or else pay with one's life. The context here is clearly reminiscent of Hobbes' classic analysis: "[D]uring the time men live without a common power to keep them all in awe, they are in that common condition which is called war; and such a war is of every man against every man."[14] All this prompts the conceptual questions of why Colombia never developed a Leviathan or a strong state that has the capacity to monopolize the use of force, and why the country has not developed effective conflict resolution mechanisms. Let us proceed to address these questions.

Fragmentation

While evidence of fragmentation in Colombia is clear enough,[15] the intriguing debate lingers as to its causes. Among the key factors are the historical absence of a shared identity, the postindependence contours of the country's notorious party politics, geographic obstacles, assorted economic elements, as well as epistemological themes. It is the synergy among these factors that has promoted Colombia's volatile cocktail of centrifugal forces. To begin, it was noted above that what is now Colombia was populated by feuding indigenous tribes during the precolonial period. This stands in contrast, for example, to the Aztec and Maya civilizations that have provided a strong historical precedent of unity and common identity in Mexico. Colombians, then, cannot look back to a period of shared identity that could help offset the profound dispersion of power in the country.

Colombians united temporarily under Bolívar, along with what is now Venezuela and Ecuador, to defeat the Spanish in 1819.[16] But factionalism became rife, especially between the Venezuelan-born Simón Bolívar and his vice president, Colombia's Francisco de Paula Santander, a phenomenon augmented by rivalry between the cities of Bogotá and Caracas. Colombian elites preferred localized power and the self-aggrandizement it provided to the prospect of centralized power under Bolívar's Gran Colombia project. With the Spanish gone and the Bolivarian project in shambles, Colombia's Liberal and Conservative parties embarked on a protracted contest to extinguish one another. Successful conflict resolution mechanisms were never established or sought. Bolívar, who died in Santa Marta, Colombia, lamented on his deathbed that "America is ungovernable."[17] While this proved to be an overstatement with regard to Latin America as a whole, it was a remarkably accurate prediction of Colombian politics over the next two centuries.

In terms of class, Colombia's Liberals represented agro-export and mercantile interests, while the Conservatives comprised the local agrarian and landed elite.[18] Conservatives predominated in former colonial centers, while Liberals represented the upstarts from peripheral regions that grew in economic significance during the postcolonial period. Neither party proved capable of achieving economic or political hegemony, either in straight coercive terms or with regard to the Gramscian conception of popular consent mixed with coercion. The fact that neither party was able to defeat the other decisively was one of the factors that contributed to both fragmentation and endless violence. Rather than working to create a centralized state, then, the Liberals and Conservatives behaved as competing and exclusive governments, hoping in vein that the next civil war would provide them with a conclusive victory over the other. Nine civil wars marred Colombia's history from 1830 to 1902.

Geographical barriers underpinned political fragmentation. Three ranges of the steep Andean Mountains presented huge obstacles for travel. This retarded the construction of roads and railways that could otherwise have assisted in connecting and uniting the country. Riverine travel was highly hazardous. During much of the nineteenth century and into the first part of the twentieth century, to arrive in Bogotá from the Caribbean, for example, one would have to embark on a trek of up to two months along the sometimes perilous Magdalena River, marked in parts by water too shallow for smooth sailing and in other areas by treacherous rapids. Insect-borne disease was rife on this sweltering journey. Once disembarking at the port near Honda, one had to climb more than 8,000 feet up the sheer Andean Mountains to reach Bogotá. Such geographical obstacles encouraged Colombia's towns to be largely self-sufficient, with little trade between regions. Within the predicament of necessary self-sufficiency, each town often produced the same things, further reducing prospects for national trade.[19] Rather than uniting into a modern nation-state, Colombian towns remained dispersed and isolated. Political rivalry between towns was the result.

Related to this is the failure of a united peasant or working-class to develop in Colombia. One reason for the lack of strong class solidarity was the physical isolation that tended to limit prospects for common political organization. Late urbanization contributed to this phenomenon. While potent unions did indeed develop, these tended to be politically and ideologically divided, as we shall see. Besides physical isolation and late urbanization, other factors contributed to the lack of a strong revolutionary class. Colombia's coffee industry, for example, worked against peasant solidarity by placating individual growers through provisions of small parcels of land and by promoting intense competition among them.[20] In addition, Colombian party politics and related clientelism absorbed peasants and other workers into binary party rivalry in a fashion that inhibited a wider class consciousness.

Epistemological factors also contributed to Colombia's notorious fragmentation. Spanish colonialism promoted a premodern system of thought. This meant, among other features, nonsecular politics, political space conceived in terms of rival city-states, and feudal economic relations as manifested through the *encomienda* system (colonial sharecropping). During the Bolivarian era, for example, even classical works that laid the foundation for modernity were unwelcome in Colombia, with Bentham's *Principals of Universal Legislation* banned as a university text in 1826 and afterward.[21] Further, Bolívar himself strongly embraced premodern ideas, as evidenced by his endorsement of a fusion between the church and state—a notion perpetuated by Colombia's Conservative Party in its strong alliance with the Catholic Church during the post-Bolivarian period. It was not until well into the twentieth century that modern ideas began to appear in Colombia with any semblance of vitality, such as the notion of secular politics and the importance of an industrialized economy. Further, as Gabriel García Márquez observed, "[O]n April 9, 1948, the twentieth century began in Colombia," following the assassination of the populist Jorge Eliécer Gaitán and the stark realization of Colombia's place within globalized class conflict.[22] But many modern ideas never took root, including the notion of balance and equilibrium, linear progress, and especially the conception of a nation-state complete with a Leviathan and conflict resolution mechanisms. There are many illustrations of this.

Colombia's geographical barriers and its relative isolation from the world economy that was centered in the North meant that ideas associated with modernity were not transmitted in any significant way to the country. That is, global economic contact helped to promote modern ideas. As we shall see, the Mexican case is illustrative of this, largely due to its location next to the United States. By contrast, Colombia remained quite isolated. Flashes of such contact appeared in the late 1800s, and increasingly during Colombia's insertion into the global economy during the 1920s and beyond. For example, the *Escuela Nacional de Minas* (The National University of Metallurgy) opened in Medellín in 1880, incorporating modern scientific ideas as foreign companies expressed interest in Colombian mines.[23] The booming coffee economy, and mounting U.S. investments in the petroleum and extractive sectors beginning in the 1920s, also catalyzed Colombia's global contact and exposure to modern ideas. The coffee economy, in particular, provided investment capital that promoted the industrial sector, which in turn underpinned urbanization—two key features of modernity.[24] But the pace was sluggish. The industrial sector accounted for 8.9 percent of Colombia's GNP in 1930, rising to 16.5 percent in 1945.[25] All this contributed to an odd patchwork of premodern and modern ideas, with neither system of thought predominating.

If these are among the reasons for the amplified fragmentation of Colombian politics, there are also some noteworthy effects. First, violence has been rife in the absence of a centralized state with a monopoly on the use of force—or a Leviathan in the words of Hobbes. Second, in the absence of a strong and centralized state, security historically has been privatized and dispersed. Examples include the private armies of *encomiendas* that were used to settle local accounts and that were also employed in interparty warfare, the development of peasant and community defense organizations, the proliferation of forces hired to protect a wide assortment of economic enterprises, the private forces of criminal syndicates, as well as a slew of other manifestations. Third, in the context of a state that has been historically weak, illegitimate, or even completely absent in many regions, economic enterprise has often operated totally outside government structures. This phenomenon would count as contraband in a strong and functional nation-state, but has not been viewed as such by many Colombians. In fact, an extensive contraband market has been documented in Colombia since colonial times, with the Caribbean Coast serving as a notorious corridor for such operations. This is the historical context of the booming industry of narcotrafficking. Fourth, in the context of virtually continuous warfare and the lack of a Leviathan, population displacement represents an important historical phenomenon, a point that will be developed more fully below. Finally, fragmentation, a weak state, warfare, and violence have conspired to prevent the emergence of a strong civil society in Colombia.

RMAs and Security for Whom? Northern Influences on Colombian Security

While domestic influences will be considered below, it is worth emphasizing that external forces historically have had a heavy hand in shaping Colombia's strategic affairs. Indeed, it is difficult to point to a history of homegrown Colombian RMAs, as we can in the Mexican case. In Colombia, neither the state nor civil society has been sufficiently strong and centralized to implement such a project. To the extent that RMAs have existed in the country, they have been externally defined to a significant degree, as exemplified by the implementation of the U.S.-concocted PC. While it may be difficult to apply the concept of homespun RMAs in an historical sense to Colombia, as one can in the Mexican case, there have been clearly definable ruptures in its strategic affairs. Let us consider some of these.

Certainly, the conquest represents an externally defined RMA with regard to Colombia—given its sweeping redefinition of power relations, strategic space, economic production, epistemology, culture, religion, language, and so on. These were driven by Spanish security interests in Nueva Granada, which centered on suppressing the local population, in order to serve Spain's twin economic interests of precious metal extraction and tropical agricultural production. While independence represented an important rupture for Colombia, many key Spanish influences remained, in terms of a feudal economic system, the fusion of church and state, language, and so on.

A fragmented Colombia was more or less left to its own devices during the postcolonial period, until about 1903. That year marked the introduction of U.S. hegemonic interests as a key determinant of Colombian security issues, a process that combined geostrategy with demands for satiating Northern consumption patterns. Within the context of the self-inflicted devastation emanating from the War of 1000 days, the United States easily wrestled Panama from Colombia in 1903 to serve its geostrategic agenda of constructing the Panama Canal. Later, with automobile production booming alongside galloping industrial growth, U.S. interests in Colombia shifted by 1915 to the country's petroleum sector—a topic that will be developed more fully below when examining the evolution of Colombia's labor unions. Important Colombian products destined for the markets of the United States and its allies during this period included petroleum, various minerals and jewels, tropical produce, and coffee.

Shifts in the world order redefined U.S. security interests in Colombia during the 1950s and 1960s. In its great contest with the Soviet Union, one that turned increasingly toward the third world during the 1950s and beyond, a central problem for Washington was to prevent the growth of Communism especially in its perceived "backyard" of Latin America. A parallel development in Colombia occurred in 1958 when the country's historically feuding Liberal and Conservative parties finally ceased warring with each other and united to fight a U.S.-shaped enemy: local Communist insurgencies. Washington's interests in the country took shape with the conception of a "national security doctrine" that relied on coercive force to contain or eliminate Communism. This package was not complemented by sufficient social welfare or developmental aid projects designed to garner popular consent—a strategically inept policy that has been accompanied by a blood-soaked pattern of failure ever since.[26]

The most significant manifestation of Washington's intervention in Colombia during the 1960s was Plan Laso. It aimed to reorganize the Colombian military to fight Communist insurgents. Related to this was the more specific Operation Marquetalia, a Colombian military adventure, which in 1964 relied on U.S. assistance to curtail peasant guerrillas, especially the FARC. Overall, Plan Laso was the biggest U.S. military aid package in Latin America until the Reagan administration's intervention in Central America during the 1980s.[27] Colombia would reclaim its title of Latin America's largest recipient of U.S. military assistance in the early 1990s.

With the United States preoccupied during the 1970s with domestic crises twinned with its defeat in Vietnam, and during the 1980s with the Reagan administration's vigorous and sometimes illegal attempts to reverse the Sandinista revolution in Central America, Colombia fell to the backburner of U.S. interests in Latin America. But with Washington's head turned, a new crisis was brewing in Colombia. It took the form of a booming drug industry—initially based on the production of marijuana before shifting to cocaine and opiates. The simple supply and demand formula meant that Colombians produced to satisfy Northern and especially American consumer habits. While the dynamics of this will be discussed in a subsequent chapter, suffice it to say here that issues surrounding American consumption contributed to burgeoning security interests in Colombia.

Thus, narco-security themes emanated from both inside and outside of Colombia. Domestically, as we shall see, the drug trade enriched to the extreme local guerrilla and paramilitary groups, and fueled the profound corruption of an already feeble state. Externally, Washington blamed Colombia for the assortment of ill effects associated with American consumption of cocaine. With the withering of the Soviet Union and its allied leftist forces in Central America, President George Bush, Sr., devised the Andean Drug War in 1989. Leaning heavily on interdiction and eradication, rather than attempting to reduce U.S. consumption, Washington militarized the Andes. In fact, by 1992 the Andean region was receiving more U.S. military assistance than any other region in Latin America. The United States provided $104.1 million in military assistance to Bolivia, Colombia, Ecuador, and Peru in federal year 1991, and $146.9 million in 1992. By the early 1990s, Colombia took the mantle from El Salvador as the largest recipient of U.S. military assistance in Latin America.[28]

Washington's perception of Colombia as a strategic problem altered once again with the inauguration of the Clinton administration in 1993. Although Colombian drug production escalated, the United States under Clinton largely ignored Colombia and the rest of Latin America except for its preoccupation with Mexico during the negotiation and implementation of the 1994 Nafta agreement. After the details for this watershed trade pact had been negotiated, the U.S. and Mexican governments shifted into panic mode with the surprise appearance on Nafta's birthday of the Zapatista guerrilla movement and the US$50 billion peso crisis of 1994–1995. Thus, Washington largely ignored Colombia during the period 1992–1998. This was a result not only of the foreign policy brush fires, but also of Washington's repugnance toward the Colombian government's apparent complicity with narcotrafficking, epitomized by the surfacing of audiocassettes that appeared to link President Samper (1994–1998) to leading Colombian drug pins.

Another dramatic shift occurred in 1998, which catapulted Colombia to the position of the world's third largest recipient of U.S. annual aid during that year. This set the stage for PC. The first obvious problem for the United States was a strategically inept decision by the fumbling Pastrana government to cede to the FARC, by then the world's largest and most powerful leftist guerrilla movement, a piece of territory about the size of Switzerland. Ostensibly, the parcel was to be used as a cease-fire negotiating region for peace talks with the government. This extraordinary development prompted Washington to bother to take a hard look at Colombia. The Clinton administration was rightfully alarmed by what it saw. Not only had the FARC grown enormously in terms of military and now territorial power, a process fed by its relation to narcotrafficking and other crimes, but other Colombian groups had ballooned as well. The leftist ELN had multiplied significantly, but perhaps the biggest growth was witnessed in the case of the right-wing paramilitaries, the AUC. These points will be developed below. Suffice it to say here that Colombia's assortment of belligerent organizations had morphed into supersubversive groups that fed off a booming illicit economy and enjoyed the elbow room afforded by a dithering government with no strategic grip.

Colombia's strategic history, then, has been marked by a series of important ruptures. These include the culmination of 9 civil wars with the devastating War of 1000 Days, as well as the descent into La Violencia in the 1940s and the establishment of the National Front in the 1950s. Another watershed appeared with the emergence of cold war guerrilla groups in the 1960s. Especially important for our purposes is the rupture that commenced in the 1980s with the establishment of a pernicious nexus in the form of crime-war-terror. This development catalyzed the formation of an externally imposed RMA under PC, as we shall see.

Social Forces and the RMA: An Historical Context

While the central focus of the next chapters will be upon the relation between the RMA, PC, and the post-2000 era, the task at hand is to provide the context for that analysis by way of a brief historical portrait of the country's principal belligerents as well as key social forces, especially the labor movement and the displaced population. Part of this discussion will include numerical estimates of Colombia's assorted groups, and so an initial disclaimer of sorts is in order. It is crucial to emphasize at the outset that such estimates may be inaccurate and laden with political objectives. All this came to light in early 2006, when over 31,000 right-wing paramilitary forces were officially "demobilized," even though the most accurate estimation of them was in the range of 20,000.[29] In that case, numerical estimates were related to a political shell game.

It can be in the interest of belligerent actors and the state to either overestimate or underestimate the membership of insurgent forces for strategic objectives. For example, if the government wishes to demonstrate that it is winning its contest with leftist revolutionaries, it is likely to report a lesser number of troops belonging to the ELN and the FARC. On the other hand, if the state is attempting to secure military and other aid from the Unites States, it may overestimate the number of rebel troops to exaggerate the necessity of funding. No one really knows the size of insurgent forces except for the insurgents themselves. Nevertheless, some numerical projections can be helpful when considering the general capacity of insurgent forces, but one must keep in mind that these represent ballpark figures at best.

The Belligerents: The FARC

This section will trace the FARC's transformation from a small agrarian-based guerrilla movement in the 1960s to a major belligerent force in the 1980s and 1990s. The group's evolution may be viewed in three phases: from its inception until 1980; from 1980 to 1990; and from 1990 until the formation of PC in 2000. According to its own documents, the FARC sprouted from an early fusion in 1950 between Liberal guerrillas and Communist "self-defense" units.[30] In 1964, Jacobo Arenas of Colombia's Communist Party joined with "resistance" forces, which included former Liberal guerrilla Manuel Marulanda (born Pedro Antonio Marín).

This formidable duo, which combined the strategic wit of Marulanda with the ideological command of Arenas, proved to be a crucial foundation of the FARC. The group existed in form by 1964, but did not officially assume its name until 1966. In terms of its support base, the FARC represented peasant farmers in the tradition of "Colombian agrarian struggles" that dated back to the 1920s.[31] Land redistribution and a more equitable division of wealth were at the center of its revolutionary objectives. FARC guerrillas concentrated in the departments of Huila, Cauca, Valle, and Tolima.[32] From its inception in 1964, the nascent FARC was met with fierce U.S. rebuttals.[33]

Perhaps the FARC's lowest ebb came within a few years of its formation. As the target of Plan Laso and related Colombian military initiatives, the FARC lost 70 percent of its armaments and a significant portion of its soldiers between 1966 and 1968.[34] To cope with that predicament, the FARC embraced at its third conference in 1968 a strategy to transform into a "mobile and very clandestine guerrilla" group[35] and refashioned itself over the next decade. By 1978, at the time of its sixth conference, FARC membership had swelled to 1,000 soldiers, and the rebels had extended their influence to both the countryside as well as urban areas.[36] In the countryside, the FARC had been pushed by the Colombian military into the interior jungles of Guaviara, Caquetá, and Putumayo. These were precisely the regions that would serve in the 1980s and beyond as the propitious economic base of coca growth and the narcotrafficking industry.

The FARC entered a new era of growth in the 1980s, when it experienced a major surge in power that was underpinned in part by the enormous economic enrichment derived from participation in narcotrafficking and other crimes. It was also bolstered by the tenure of the Sandinista revolution in Nicaragua from 1979 to 1989, which provided a glimmer of moral support and hope for revolutionaries throughout Latin America. The FARC's seventh conference in 1982 marked an official turning point for the increasingly empowered rebels. The FARC, in its own words, had now transformed into an "offensive movement," with designs for a greater presence throughout the country and in major urban areas such as Bogotá.[37] Its goals included the establishment of 48 fronts, the achievement of greater strategic adaptability, and the development of a more sophisticated means of communication to benefit economic, ideological, and military objectives.[38] During this period, then, the FARC underwent a major transformation from a small, beleaguered, Cuban-styled guerrilla movement on the run, to a major belligerent force bankrolled by transnational crime. It was during this epoch that the FARC first used the nomenclature of *Bolivariana,* harkening back to the ideas of Simón Bolívar. There was a particular focus on his nationalist dreams of forming a sweeping Latin American movement capable of resisting the imposition of Northern powers.

In 1984, the FARC embarked on a major program of political development with the creation of its political unit, the Unión Patriótica (UP). This occurred in the midst of what turned out to be failed peace negotiations with President Belisario Betancourt. The UP scored well during its electoral debut in 1986, winning 14 congressional seats and numerous positions in departments and at municipal levels. It obtained about 350,000 votes in the presidential contest. This trend toward increasing political popularity, especially in the countryside, continued into the early 1990s. Tragically, this electoral success translated into a major rupture for the country when about 3,000 to 4,000 members and supporters of the UP were assassinated by paramilitary forces between 1986 and 1992. The FARC interpreted this gruesome phenomenon to mean that the state was complicit in the assassinations through its failure to protect UP members, and also due to the impunity it afforded to paramilitary assassins.

In tandem with the devastating experience of the UP, in 1990, President Gaviria launched a vigorous military attack against the FARC at Casa Verde, killing the group's ideological cofounder, Jacobo Arenas. For the rebels, then, there appeared to be no space in "legitimate" or peaceful politics. The significance of this turning point cannot be underestimated, because it prompted the FARC to do whatever it took to transform into a major military machine designed either to topple the state or to form a parallel government in a territory it controlled militarily. While half-hearted peace talks took place in 1992, the context was not ripe for any meaningful success. During the FARC's eighth conference in 1993, the rebels planned a major offensive to be launched in the last half of the decade[39] and increased its number of fronts by 15.[40]

The first salvo of this new offensive occurred in 1996, when the FARC overtook humiliated Colombian military forces, including those stationed at bases in Las Delicias, La Carpa, and de Patascoy. The group's crowning achievement came in 1998, when the strategically inept government of Pastrana ceded to the FARC a "cease fire" zone, or *zona de distensión*. The provision of this parcel was designed to placate the FARC and to provide the basis for reestablishing negotiations. Instead, and predictably, it empowered the rebels, who now wished to preside permanently over a swelling territorial enclave. Armed confrontations between the FARC and the military grew from an average of about 150 annually during 1985–1990, to about 400 annually for most of the 1990s, jumping to about 1,200 when PC reached full tilt by 2003.[41]

By the late 1990s the FARC had mobilized an estimated 17,000 troops spread over 60 fronts.[42] It secured political support from *cocaleros* (coca growers), destitute peasants, and student radicals scattered among the country's universities. In an attempt to distance itself from what some viewed as outdated cold war Communism, the FARC increasingly began using "Bolivarian" rhetoric throughout the 1990s, essentially rhyming ideologically with dimensions of President Hugo Chávez' revolutionary platform in neighboring Venezuela. In addition to local and regional support, the FARC had some limited success at courting European NGOs and even European governments, and in 1999 hosted in its jungle *zona* the chair of the New York Stock Exchange. In the next two chapters we shall explore the FARC's place within PC and the RMA. For now, let us proceed to examine the historical context for the emergence of other insurgent forces and of key elements of civil society.

The Belligerents: The ELN

The ELN, Colombia's most important guerrilla group after the FARC, was formed in 1964. It was founded by Fabio Vásquez, with strong support from a Catholic priest named Camilo Torres. It began in San Vicente Chucurí, Santander, and officially describes itself as an "insurgent group" that blends "Marxism-Leninism" with "Revolutionary Christianity."[43] The group was inspired by the 1929 Communist uprisings, the Liberal guerrilla wars of La Violencia, the student radicals at the Universidad Industrial de Santander in Bucaramanga, and especially the Unión Sindical Obrera de la Industria del Petroleo (USO) as well as other labor unions.[44] Indeed, in 1965, Torres visited the petroleum workers' USO, perhaps the country's most radical labor union, in the oil hub of Barrancabermeja.[45] Further, while Torres lost his life during combat in 1966, the strong element of Revolutionary Christianity has continued to be a distinguishing feature of the group. Ideologically, the rebels have attempted to mix the "foquista' " ideology of Che Guevara with a Christian doctrine that views murder as the ultimate sin.

Thus, the profound conundrum for the ELN has been how to participate in guerrilla warfare in one of the most violent countries in the world while avoiding murder. In the words of its recent leader, Nicholás Rodríguez Bautista (aka Gabino), Colombia is a "violent society, but this is not a natural condition, and within the context of violent institutions, the people must respond equally to defend themselves."[46] So the goal, one that has not always been achieved, has been to fight in self-defense and to avoid offensive murder. The ELN has not been able to compete militarily on an equal footing with the FARC, given the latter's Clausewitzian reliance on force. A further distinction between the two rebel organizations is that while the FARC is composed of full-time, professional soldiers, the ELN has relied largely on a moonlighting force that divides its time between various work-a-day jobs and a commitment to the rebels.

Due to a series of military defeats at the hands of the Colombian government during the late 1960s through the 1970s—the same military program that had been launched against the FARC—the ELN dwindled to just 30 members by 1978. This shrinkage was also generated by infighting among members, especially regarding the contentious issue surrounding the use of force. By the early 1980s, the remaining members of the ELN were pressed by the military into the Magdalena Media area and into relatively forgotten regions of the country near the border with Venezuela.[47]

But fortunes changed for the better for the ELN by the mid-1980s. Especially with regard to the regions of Arauca and Norte de Santander along the Venezuelan border, the group had found itself essentially exiled into an area where the state was absent and where huge new oil reserves were suddenly discovered. A case in point is Occidental Petroleum's 1983 discovery of oil in Arauca. At the same time, the ELN concentrated its efforts on one of its original areas of support—the oil hub of Barrancabermeja in the Magdalena Media region. Against this propitious backdrop, the ELN's Gabino, along with other leaders of the group, devised a strategy for the period 1983–1986 that would "tax" oil corporations and other extractive industries. This typically entailed the threat of kidnapping or the destruction of key installations through bombings and the like.[48]

Thus, while the group officially balked at offensive murder in terms of revolutionary strategy, it still embraced violence of another sort, and remained quite distant from Gandhi's strategy of passive resistance. A gripping and sadly truthful slogan of the group, often found on the covers of its official pronouncements, is that "to the disgrace of Colombian society, the governing class will only listen to the voice of dynamite and guns."[49] Thus, the ELN launched a strategy that generally was able to resist offensive murders by relying instead on extortion, kidnapping, and bombings of key sites. The growing coffers of the rebel group financed a swelling roster of recruits. ELN troops are roughly estimated to have increased from 30 in 1978 to 800 in 1986 and 3,000 in 1995, to a peak of about 5,000 by the turn of the century.

The ideology, policies, and supporters of the ELN have remained remarkably consistent since its inception. From the beginning, ELN has called for the popular classes to seize power and establish policies related to agrarian reform, the nationalization of key industries, a popular system of credit, a national health plan, and educational reform. More recently, it has stressed the plight of Colombia's indigenous and black populations, and has also emphasized ecological issues. The ELN has condemned the dangerous and corrupting influences of narcotrafficking that blacken its rivals, including the FARC, the AUC, and components of the state.[50] It has also provided strong support to the plight of labor movements, especially regarding themes such as antiprivatization and the increasing use of cheaper, nonunion staff. Its supporters have included students, members of key labor unions (especially those in the oil industry), and disenfranchised members of the middle class who have been attracted to the group's revolutionary doctrine and its expressed repugnance to offensive violence.

The ELN sought to maximize its power in the 1990s, especially in the last few years of the decade. The group launched a series of blistering critiques regarding neoliberal policies imposed by the IMF in Colombia and throughout Latin America. A central objective for the ELN during the late 1990s was to attain a piece of territory under its control, especially after the FARC was able to convince the government to provide it with its *zona*. The ELN unleashed a series of stunning tactics designed to grab national and global headlines in an attempt to bolster the bargaining power of the group. Among these was the kidnapping of an entire planeload of people on an Avianca flight from Bucaramanga in 1999, and the kidnapping of an entire church congregation in Cali during the same year. The ELN was able to attract considerable media attention through these tactics rather than through the murderous, offensive combat that characterized the FARC's campaign. By the time the Pastrana government was willing to cede to the ELN its own cease-fire zone in April 2000, amplified criticism of the project by Colombian society, and probably clandestinely by the U.S. government, which had now concocted PC, meant that it never reached fruition.

By the end of the 1990s, the ELN exhibited both similarities with, and distinctions from, other Colombian rebels. Like the FARC and the AUC, the ELN was largely self-sufficient economically and did not succumb, as Central American revolutionary groups did, to the swan song of the Soviet Union. It has shared with the FARC lessons extracted from the dreadful experience of the UP. Like the FARC, it viewed the 1990s as a quest for power maximization fueled by criminal revenues. But the similarities among these groups stop there. Unlike either the FARC or the AUC, the ELN did not participate heavily in the narcotrafficking industry. Its official pronouncements have been critical of the illicit drug trade and its relation to the country's vicious Dirty War.[51] It has not launched campaigns of unbridled carnage that have characterized the actions of the FARC and the AUC.

Although the ELN occasionally relied on a weak and desperate alliance with the FARC during the PC era, as we shall see, it never achieved any significant measure of harmony with the FARC. With regard to its other rival, the ELN throughout the late 1990s bitterly condemned the AUC, which has fought strenuously against the ELN and was chiefly responsible for ELN's near demise in the first decade of the new millennium. The ELN has regularly accused the AUC of representing the Colombian state and sectors of the economic elite within the context of a Dirty War.[52] The ELN reached the height of its power in the 1990s and, as we shall see in subsequent chapters, has faced a steady decline in the new century at the hands of the AUC, the FARC, and the Colombian military.

The Belligerents: The AUC

> The most important thing is that from now on we'll be fighting only for power.[53]
>
> *Gabriel García Márquez,* One Hundred Years of Solitude

The AUC did not become a major political and military force until the 1990s. But the paramilitaries have roots dating back to 1965, when President Guillermo León Valencia enacted legislation permitting self-defense groups to fight against leftist insurgent movements.[54] Their outline became clearer in the late 1970s and into the 1980s, when the paramilitaries began to serve a number of social functions. First, they acted as the illicit security wing for wealthy and sometimes even middle-class landowners. This was manifested in a number of ways. For example, when the Colombian government started talking seriously about land reform in the late 1970s as a method of coping with leftist forces, paramilitary units assembled and attacked leftist groups.[55] Paramilitaries also protected those who came under the threat of kidnapping and extortion from guerrilla groups such as the FARC.

In February 2005, the paramilitaries traced their definitive origin to February 1978, when self-defense groups organized to repel an attack by the FARC in San Luis, Antioquia.[56] Another inflection occurred in 1981, with the debut of Muerte a Secuestradores (MAS—Death to Kidnappers). A similar group was the Muerte a Revolucionarios de Nordeste (MAN—Death to Revolutionaries of the Northeast), which operated in Antioquia and protected the interests of major mining corporations.[57]

As the 1980s progressed, and as important narcotraffickers became proprietors of large landholdings, right-wing paramilitaries became financed by drug barons to act as their security apparatus. Overall, the fortification of the paramilitaries in the 1980s was supported by large landholders, narcotraffickers, businesspeople, extractive corporations, transporters, agroexporters, and so on.[58] In general, the paramilitaries have represented the class interests of often illicit national capital, in addition to their allegiance to transnational capital and the strategic interests of the United States. One observer suggested that all this boiled down to the feudal practice of "exacting tribute in exchange for the regulation of public order and the protection of the market."[59] The paramilitaries have seen themselves as defenders of the ideological roots associated with the existing system, in contrast to the aims of leftist groups such as the FARC and the ELN that struggle for revolutionary and systemic change.[60] In the 1980s, these self-defense units admired Ronald Reagan and his illegal creation of the Contras in Nicaragua—paramilitary forces working at the behest of the U.S. government to combat the revolutionary government of the Sandinistas. Similarly, current critics of the paramilitaries view them as clandestine auxiliaries of the Colombian and perhaps U.S. governments, a point that will be developed in subsequent chapters.

The AUC was officially founded in 1994, and was concentrated initially in the areas of Cordoba and Urubá. Unlike the FARC and the ELN, which are unified, the AUC is an umbrella organization of numerous and sometimes unruly paramilitary groups. In 1994 the government sponsored "Conviver" self-defense units by arming peasants, and there is a likely correlation between this and the rise of paramilitary squads beginning in that year. Until 2000, the AUC comprised the following loosely knit groups: Autodefensas Campesinas de los Llanos Orientales, las Autodefensas Campesinas de Santander y Sur del Cesar, Autodefensas Campesinas de Magdalena Medio, and Autodefensas Campesinas de Cordoba y Urubá. Overall, the AUC was the fastest growing subversive group in Colombia during the 1990s. Its growth correlated with a steep increase in ferocious human rights abuses. U.S. statistics indicate that human rights violations attributed to the AUC rose 100 percent between 1995 and 1996. Paramilitaries were responsible for 69 percent of assassinations in 1997, and were the cause of about 35 percent of the country's population displacement by 2000.[61]

There were a number of factors at play that contributed to the growth of the AUC up to the year 2000, the figures of which are documented in Table 2.1.

Table 2.1 Growth of paramilitary troops by year

Year	*Troops*
1986	93
1987	650
1989	1500
1990	1800
1991	1150
1992	850
1993	1200
1994	2150
1995	2800
1996	3400
1997	3800
1998	4500
1999	5915
2000	8150

Source: Ejército Nacional, Ministerio de Defensa, "Los Grupos Ilegales de Autodefensa de Colombia," Colombia, December 2000.

One of these factors is that Carlos Castaño came to the helm of the group in 1997, after the death of his narcotrafficking brother and cofounder of the rebels, Vicente. Carlos Castaño endeavored to unite the dispersed right-wing paramilitaries and to present this unified group to the Colombian public as both a major military force and an articulate political voice. Regarding other factors, an insightful Colombian analyst has suggested that surges of paramilitary growth have coincided with the initiation of serious peace negotiations between the government and leftist guerrillas such as the FARC.[62] This was most evident in the early 1980s, when the Colombian government was negotiating with the FARC between 1982 and 1986, during the UP era of the late 1980s and early 1990s, and during the period surrounding the establishment in 1998 of the government-condoned *zona de distensión* for the FARC.

Additional debates flow from that premise. Was periodic growth of the AUC a violent response by the business class to signal its failure to accept government policies that were too friendly with the Left? Were components of the AUC working in concert with the Colombian military while the government was negotiating with the FARC, playing out a hard-cop/soft-cop arrangement? Presumably, there is some truth in both of these suggestions. A third factor is also important. Colombia's Ministry of Defense suggests that the growth of the AUC during the late 1990s was directly linked to increased coca growth—that is, the AUC has essentially represented the security wing of a burgeoning narcotrafficking industry. While coca growth in Colombia increased from 80,000 hectares in 1996 to 130,000 in 1999, the number of paramilitaries grew from 3,400 to 5,915 during the same period.[63] In the new millennium, the AUC represents an important component of the RMA in Colombia—a point that will be developed in the next two chapters.

Social Forces and Civil Society: The Case of Labor

> But my most striking memory from that time was the swift passage of the superintendent of the banana company in a luxurious open car, beside a woman with long golden hair that blew in the wind, and a German shepherd sitting like a king in the seat of honor.[64]
>
> *Gabriel García Márquez,* Living to Tell the Tale

Here we shift from a consideration of Colombia's insurgent forces to social actors engaged in peaceful struggle. The labor movement is important to this study because unions historically have been among the best-organized components of Colombia's notoriously weak civil society.[65] In the country's highly exclusionary social context of violence mixed with terror, union members historically have mustered considerable bravery and fortitude to represent the basic interests of workers. Against such a backdrop, their mere existence is truly remarkable. Although union membership has contracted steadily since the mid-1980s to the beat of global neoliberal restructuring, labor continues to be a strong voice regarding themes that lie at the heart of the Colombian political imbroglio. This section will briefly trace the history of the Colombian labor movement, and will set the stage for an analysis in subsequent chapters regarding the unions' role at the intersection of the social struggle and the RMA.

Colombia's first workers' organizations began as *sociedades de mutuo auxilio,* or mutual support societies. These appeared in Manizales in 1889, Bucaramanga in 1890, Giradot in 1908, and Sansón in 1909.[66] The first major unions formed near the turn of the century, with the appearance of the Union de Industriales y Obreros in 1904 and the Union Colombia Obrera in 1913. A key era of union formation and of major political strikes occurred during 1918–1930. By this time, the Soviet Union emerged as a major force, socialist parties were developing in many parts of the world, and the labor movement was feeling the wind in its sails. With regard to Colombia, unions were forming to meet the challenge associated with increased U.S. investment that appeared during this time. This epoch was marked by a wave of huge strikes in 1918 at Colombia's ports in Cartagena, Santa Marta, and Barranquilla. Troops were called in to suppress these strikes, which represented the first salvo of a violently confrontational relationship between government forces and labor. By 1919 there were 26 unions in the country.[67] In terms of party politics, the Liberals drew most of their support from urban areas, and hence began to form a relationship with the labor movement during this period. This was especially the case with regard to the socialist component of the party, which incorporated leftist intellectuals such as Jorge Eliécer Gaitán in 1923.[68]

The two most serious sets of strikes occurred in the petroleum and banana sectors. With regard to oil workers, the most important labor action occurred in 1924 and 1927. The issues were similar. Basic hygiene and health concerns topped the list, followed by requests for an eight-hour workday, rest on Sunday, and so on. For example, 1,023 of 2,838 oil workers in the tropical city of Barrancabermeja became gravely ill in 1923–1924, leading to basic worker demands concerning the placement of screens on the windows of company housing to protect them from preventable insect-borne diseases. When oil workers took to the streets of Barrancabermeja in 1924, troops moved in and forced the expulsion of some 1,200 workers from the city.[69]

An infamous strike occurred in 1928 among banana workers at the United Fruit Company's operation near Santa Marta. Worker demands—many of which remain relevant almost a century later—included hygienic working conditions, better wages, and freedom from purchasing goods only at the company store. They were also calling for the elimination of a "contractor" system whereby the company could claim that as the vast majority of workers were subcontracted, they were not actually employees of United Fruit. By December 1928, some 25,000 workers stopped cutting bananas, triggering a notorious massacre of workers that was accompanied by a wave of assassinations.[70] The petroleum and banana industries were bellwethers for labor issues in other sectors. There were significant strikes, for example, in Antioquia's gold mining companies in 1919 and 1926.[71]

Some broad points follow from this. First, both the petroleum and banana sectors remain central to current union politics in Colombia, and represent key social forces within the context of warfare and the RMA. Second, the 1920s initiated an enduring scenario whereby the labor movement was generally viewed by government and business leaders as a subversive socialist force that occasionally needed to be quelled through violence. Third, a legacy was born whereby members of labor organizations, and especially union leaders, became targets of systematic violence and assassination. Further, some observers sympathetic to labor suggest that the general inability of unions to realize their demands and the related lack of conflict resolution mechanisms have promoted violence through a tendency for labor demonstrations to express popular frustration by evolving into riots.[72]

By the beginning of the 1930s, Colombia's era of "heroic unionism" that characterized the 1918–1930 epoch gave way to a mixed bag of victories and defeats for Colombian labor. Fragmentation and disunity, which have plagued Colombian politics in general, grew stronger among unions. Some of this had to do with debates concerning the allegiance between unions and Communist organizations, among other matters.[73] On the positive side of the ledger, important legislation was established that officially recognized organized labor in 1931, and an eight-hour workday was codified in 1934. The Confederación Sindical de Colombia (Colombian Confederation of Labor Unions), an umbrella organization formed during 1936–1938, evolved into the Confederación de Trabajadores de Colombia (CTC—Colombian Confederation of Workers) in 1943. The CTC was strongly allied with the Liberal Party, and its formation coincided with a split among labor regarding support for the Liberals versus the Communists. It also heralded an era of less frequent strikes, partly due to the CTC's investment of energy into the election of President López in 1942. The 1930s to the early 1940s, then, witnessed an institutionalization of labor, which enhanced its role as the largest and most important political organization within Colombia's weak civil society.

The National Front was established during 1958–1974 as the Liberals and Conservatives agreed to halt more than a century of bitter feuding and mutual political exclusion. The two parties fused their interests regarding the perpetuation of a capitalist political economy, and cooperated with regard to their common animosity toward the Left. Labor movements, especially those sympathetic to socialism, came under increasing suspicion and surveillance by the government during this period. Overall, for the government and the military, Left-leaning unions were counted as among the enemy.[74]

The 1960s and 1970s marked important developments for labor. By 1966, legislation permitted government workers to strike legally, but also limited the number of annual legal strike days to 43. Union membership levels were significant. For example, 47 percent of finance workers were unionized in 1974, compared to 46 percent of industrial workers. Figures for unionized labor in other sectors during that year included 43 percent for transportation and communications, 41 percent for water and gas, 21 percent for mining, 8 percent for construction, and 7 percent for agriculture.[75] The number of strikes increased from 576 during the period 1962–1971 to 669 during 1971–1980. During the same periods, the number of strikers increased from 943,000 to 1,786,000.[76] Many view the 1977 national strike as a turning point in Colombian labor history, as its general failure precipitated a debilitation of the labor movement. Perhaps the situation of the petroleum union USO epitomized the strike results: no worker demands were met, 217 workers were dismissed from their jobs, and another 45 were detained by authorities for interrogation.[77] More generally, unions and workers became increasingly fragmented ideologically, with pronounced ideological divisions among Communists, socialists, anarchists, liberals, social democrats, and conservatives.[78]

The context for labor altered dramatically in the 1980s. The intensification of Colombia's armed conflict coincided with an environment of increasing insecurity for union members, in the form of assassinations, massacres, harassment, and repression. This atmosphere of extreme insecurity is one of the key reasons for the weakness of the country's civil society—freedom of expression in Colombia has often been threatened by political assassination. More than 3,800 union leaders and members were murdered between 1985 and 2002.[79] Beyond the unspeakable carnage directed against those struggling for basic workers' rights, a period of layoffs further debilitated unions after 1981.

During the 1990s, a trend toward privatization, as well as restructuring policies imposed by the IMF, helped define issues for Colombia's beleaguered union movement. While workers' issues were clearer and more pressing than ever, the roster of Colombian organized labor dwindled due to an increased use of contract workers and the privatization of national industries. Union membership declined from 844,166 members in 1994 (.0538 percent of the actively employed population) to 837,521 in 2002 (.0428 percent).[80] Clearly, labor is among the components of Colombian civil society that is most affected by armed conflict, and in subsequent chapters we shall situate this within the context of PC and the RMA.

Social Forces and Civil Society: The Case of Colombia's Displaced Population

No component of Colombian society has been more affected by the country's virtually interminable warfare than the displaced population. Typically, their plight has meant the loss of identity, livelihood, homes and land, a support network of relatives and friends, and so on. Overall, displacement spells the abrupt absence of social, psychological, and economic anchors. This, in turn, promotes a profound sense of insecurity. The issue of displacement has played out differently in distinct historical eras. Further, in the past and present, warfare has not been the only cause of displacement. Historically, for example, peasants regularly sought to escape the feudal confines of the hacienda for a greater degree of economic and political freedom. The plight of economic refugees persists into the twenty-first century, and sometimes the distinction between warfare and economic destitution as a cause of displacement is blurred.

While displacement has accompanied war, its levels have waxed and waned depending on the intensity of combat and other factors.[81] For example, during periods of horrific violence resulting from battles between the Liberals and Conservatives, such as the War of 1000 days of 1899–1902, and especially during La Violencia of 1948–1958, displacement was a predictable phenomenon. There was an estimated 2,003,600 persons displaced during La Violencia.[82] During that era, many Liberal peasants fled to Guaviare, Caquetá, and Norte de Santander, where leftist guerrilla groups would later form.[83] Certain regions of the country have a longer history of displacement, such as the Magdalena Media, while others have experienced significant displacement only recently.[84] It is important to point out that during La Violencia and earlier eras, cities were relatively safe havens for the displaced to settle.[85] But this has not been the case since the 1980s, when they found sparser prospects for employment and little or no social services, and were stigmatized as thieves and criminals.

The 1985–2000 phase can be divided into two eras: 1985–1995 and 1996–2000, as Table 2.2 demonstrates. Data on the number of displaced was not collected and organized in any significant way until the formation in 1985 of the Codhes, which emerged as Colombia's leading NGO for the displaced and the most respected authority on the subject.[86] There are a number of relevant methodological concerns surrounding the issue of who counts as a displaced person. For example, the government did not officially recognize violence as a cause of displacement until 1994. Further, Codhes presents both annual and global figures, and these are estimates based on the assumption that only about 30 percent of the displaced population actually registers with an agency (such as the Codhes, the church, the Red Cross, or the government's Red de Solidaridad Social). Most do not register for fear of reprisal from those they are fleeing, or due to the heavily negative social stigma attached to displacement. Codhes' global figure of approximately 3.4 million displaced persons between 1985 and 2005 led the assistant UN high commissioner for refugees to declare the Colombian humanitarian crisis to be the worst in the world outside Africa.[87] In contrast to Codhes' global statistics, the government has disseminated only annual figures of displacement, and typically counts a person as displaced only for a matter of months after initial displacement.[88] Further, the

Table 2.2 Number of displaced in Colombia, 1985–2000

Phase one:	1985	27,000
	1986	36,000
	1987	59,000
	1988	105,000
	1989	119,000
	1990	77,000
	1991	111,000
	1992	64,000
	1993	45,000
	1994	78,000
	1995	89,000
Phase two:	1996	181,000
	1997	257,000
	1998	308,000
	1999	288,000
	2000	317,000

Sources: Consultoría para los Derechos Humanos y el Desplazamiento (Codhes), *Un Pais que huye: Desplazamiento y violencia* (Bogotá: Codhes, 2003); and Codhes, "Número de Personas Desplazadas por Departamento por Trimestre, 1999–2004," pamphlet, 31 January 2005.

government discounts those displaced from regions where fumigation of illicit crops has occurred.

During the first phase, from 1980 through 1995, the annual level of displaced persons tended to rise steadily and was an important indication of the growing severity of war. There are a number of distinct phases within that period. For example, the displaced population grew steadily from 1985 until 1987, as the battle escalated between the army and the left-wing guerrillas. There was also growing social violence emanating from pre-paramilitary groups such as the MAS. As Table 2.2 demonstrates, displacement rose sharply during the years 1988 to 1991, largely resulting from the terror associated with the assassination of thousands of UP supporters. While the immediate post-UP era of 1992 to 1995 meant a slight decrease from the 1988–1992 crescendo, these numbers mostly remained higher than pre-1988 levels, and grew steadily until a political explosion in 1996.

The 1996–2000 period witnessed a huge upsurge in the displaced population due to intensifying warfare. The quest for territory and resources, and related attempts at power maximization among the belligerents and narcotraffickers, together with an ever weaker state, contributed to horrific levels of displacement. In terms of responsibility for displacement, the paramilitaries take the lead according to statistics compiled by Codhes. Their responsibility was measured at 35 percent in 1995, 33 in 1996, 54 in 1997, and 47 in 1998. For the guerrillas, the figures were 26, 29, 29, and 35 percent in the same years, respectively, with the military responsible for the rest.[89]

Finally, it is important to reiterate that in Colombia the displaced population is the social force that is perhaps most negatively affected by increasing levels of violence and warfare since 1980. This has had a multitude of abysmal effects. Chief among these are family breakup, with women becoming the heads of destitute households in a strange land. Poverty is also a product of the abandonment of land and houses among this population, who are forced to leave quickly with only what they can carry. For example, 3 million hectares of land (many with houses) were abandoned between 1996 and 1999.[90] Children, who make up approximately 75 percent of the displaced,[91] are uprooted from school in many cases, and must work in ghastly conditions to support their families. Some have felt driven to crime to support themselves in a climate of desperation and economic depression, a development that fuels the stigma of criminality attached to the displaced. Colombia's indigenous and blacks have suffered disproportionately.[92] The indigenous are often displaced from lands rich in oil or minerals, as exemplified by the situation in Arauca. As we shall see, blacks from the Pacific region of the country have been displaced from their homeland because it has become a theater of warfare between the paramilitaries and left-wing guerrillas, as well as the site of mega-development projects such as palm oil plantations that proliferate at the expense of local farms.[93] The subsequent chapters will trace the plight of the displaced in the context of PC and the RMA.

* * *

This chapter has provided an historical backdrop from which to consider social forces in relation to PC and the RMA. We observed the profundity of violence and fragmentation in the country, and have traced an historical series of strategic ruptures. We shall see in the following two chapters that the RMA associated with PC was implemented by the United States for a variety of intriguing reasons. Chapter 3 discusses the strategic package of PC, with a special emphasis on features of the RMA including asymmetry, complexity, and surveillance-intelligence. These will be examined in relation to belligerent forces as well as to components of Colombia's civil society—especially the labor movement and the displaced population. Chapter 4 will focus on social forces and the privatization of warfare.

3

Plan Colombia

Colombia's pronounced violence and fragmentation, along with an intriguing patchwork of epistemological considerations, continue to represent essential themes in the country's imbroglio. Now that Colombia's roster of belligerents has been introduced, as were the social forces of labor and the displaced, we can proceed to view them in relation to basic elements of Plan Colombia (PC). Key aspects of the RMA are examined with regard to PC, such as the discourse on terror, asymmetry and complexity. There is a special emphasis here on the crucial components of ultra-surveillance and intelligence. We shall see how these various elements of the RMA work together with a synergetic effect to explain many crucial aspects of strategic change in Colombia.

Plan Colombia

PC was unveiled by Colombian authorities in 1999 and approved by the U.S. Congress in July 2000. At the outset, it envisaged $7.5 billion in aid to Colombia until 2006, although only about $4.7 billion was actually allotted during that period. At least 75 percent of this has turned out to be military or police aid, with much of it dedicated to the "drug war." The rest has been devoted to institutional programs, especially in the judicial sector, and to a lesser extent to social programs.[1] Washington had expected contributing funds from other donors, but in the main these failed to materialize. A case in point concerned expected funding from the European Union, whose skeptical members hesitated to support what many viewed as a militarily heavy-handed project driven by the Americans.

To place in perspective the funding associated with PC, let us consider U.S. aid patterns to Colombia since the 1990s. Table 3.1 indicates the declining attention Washington afforded to Colombia, particularly beginning with the inauguration of the Clinton administration in 1993. U.S. aid ebbed to a trickle in the 1994–1996 period, for reasons discussed in the historical overview presented in the previous chapter. As Table 3.2 demonstrates, U.S. antinarcotics assistance almost tripled from $109.38 million in 1998, just after the FARC received its *zona,* to $294 million in 1999, and then ballooned to an astonishing $756.76 million in 2000, the inaugural year of PC. Paralleling this increase in aid was a substantial boost in the

Table 3.1 U.S. military aid to Colombia, 1990–1996 (millions, US$)

	1990	1991	1992	1993	1994	1995	1996
IMET[a]	1.54	2.76	2.31	2.641	0.9	0.588	0.95
MAP[b]	0	16.07	8.40	4.096	0	0	0
FMF[c]	69.731	47	47	27	7.7	10	n/a
Total	71.271	65.83	57.71	33.737	8.6	10.588	0.95

Source: U.S. Department of Defense, Defense Security Cooperation Agency, 2001, www.dsca.mil/programs/Comptroller/2001_FACTS/.
[a]IMET is International Military Education and Training.
[b]MAP is Military Assistance Program.
[c]FMF is Foreign Military Financing.

Table 3.2 U.S. military aid to Colombia, 1997–2007 (millions, US$)

Year	NCA[a]	FMF, IMET Sec. 1033, EDA[b]	Antiterrorism assistance	Total
1997	58.47	30.09	0	88.56
1998	109.38	3.06	0	112.44
1999	294	14.81	0	308.81
2000	756.76	8.56	0	765.32
2001	196.39	28.29	n/a	224.68
2002	339.19	7.55	25	371.74
2003	567	34.87	3.28	605.15
2004	473.5	127.89	0	601.39
2005	463	100.9	5.1	569
2006 (estimate)	465	90.7	5.3	561
2007 (request)	465	91.7	3.1	559.8

Sources: U.S. Department of State 2005, 2006; U.S. Department of Defense 2001; Center for International Policy 2004, 2006.[2]
[a]NCA, Narcotics control assistance, includes the U.S. State Department's International Narcotics Control, Presidential "Emergency Drawdowns," Section 1044 of National Defense Authorization Act, which provides additional support for counterdrug activities, and discretionary funds from the Office of National Drug Control Policy.
[b]FMF is Foreign Military Financing, IMET is International Military Education and Training. This figure also includes Section 1033 of the National Defense Authorization Act, sometimes called a "Riverine Program" as it focuses on interdiction of drugs along river routes. Also included in this figure is EDA, Excess Defense Articles.

realm of military materiel and training. This was further bolstered by the provision since 2002 of antiterror support in the wake of the 9/11 attacks.

Within the context of the post-9/11 environment and the subsequent invasion of Iraq, Colombia as a U.S. strategic concern lost some of its luster but remained significant. In terms of global rankings for U.S. military assistance, Colombia fell from third to seventh by the end of 2005. In what would become a trend beginning in 2004, U.S. aid to Iraq shot up to $18.44 billion, dwarfing the amount devoted to other countries.[3] Thus, while Colombia's strategic significance did not compare in intensity to U.S. homeland security or to postinvasion projects in Iraq, it nevertheless remained on Washington's shortlist of global priorities. This is substantiated by the continuation of PC until 2006 and the Bush administration's attempt in 2007 to provide additional support over the next few years. The Colombian imbroglio has represented an important experimental theater of warfare within the context of hemispheric affairs. It has exhibited many of the same features of the RMA that Washington has faced with regard to its Middle Eastern foes, although with distinct manifestations.

PC, as a Washington-designed "RMA from above," addressed three interrelated problems for the United States. PC was intended to subdue the FARC, to protect American economic interests particularly in the mining and oil sectors of Colombia, and to promote U.S. strategic and economic interests in the Andean region as a whole. A broader but equally significant problem for the Bush administration was how to achieve these goals during a limited time frame, and at the same time, to avoid bad press and public scrutiny in the process.

War, Terror, and Plan Colombia

Terror is the amplification of fear for larger audiences, with the terrorist act itself serving as a public spectacle designed to disseminate a political message throughout society. There exists great variation among terrorists, as evidenced by historical groups such as the Thugs, Assassins, and the Zealots.[4] More recently, one notes the distinction between holy terror, featuring suicide bombers, versus Western remote-control terror, wherein the person detonating the bomb wishes to continue pursuing earthly delights. Within Latin America, terror has been employed by both state and nonstate actors.[5] The definition of terror cited above concentrates on the act of terror—the exaggerated use of fear to achieve a political goal—rather than on the actor. By contrast, governments typically prefer to define the term in a manner that erases any consideration of state responsibility for such acts. Hence there is a propensity for governmental figures to define terror as a violent tactic of nonstate insurgents acting within the context of asymmetric warfare.[6]

Beyond the issue of terrorist acts and actors, the discourse on terror can represent an important instrument of power. Clearly, the Bush administration discovered the relevance of terror to PC. Almost overnight, Colombia's assortment of subversive groups—especially the FARC and the ELN—shifted from their casting as a "guerrilla group" engaged in a "civil conflict," as was the case in the 1990s through the summer of 2001, to global "terrorists" likened to more than 30 other such groups on a newly composed list of U.S. enemies. The first hint of this came with comments in October 2001 by James Mack, U.S. deputy assistant secretary of state for international and law enforcement affairs, who pointed to a "nexus between terrorism and organized crime" in addition to noting that "many of the skills and types of equipment needed to attack organized crime are applicable to combating terrorism."[7] The clearest statements in this regard were made by President Bush himself in April 2002 during a meeting in Washington with Colombian president Andrés Pastrana. Bush praised Pastrana for fighting "terrorism in his country," and underscored a fundamental change in U.S. policy toward Colombia, when he commented that he and Pastrana "had a good discussion about a variety of issues about how to change the focus of our strategy from counter-narcotics to counter-terrorism."[8]

The discourse on terror has meant far more than merely a change in nomenclature. It permitted the United States to be more transparent with regard to the enemies it defined in Colombia, and to present a more overt strategy for dealing with them. Prior to 9/11, PC was largely portrayed as an antinarcotics campaign. Close observers realized that it was far more than that, as it principally attempted to eradicate coca crops in the south of Colombia, from which the FARC benefited, and also included a refashioning of the Colombian military that was clearly aimed at the guerrillas. But since 9/11, and with the FARC and ELN recast as terrorists, the United States became completely open about its attempts to confront the FARC militarily. This transparency permitted Washington to intensify PC's focus on antiguerrilla activity, for example, by utilizing part of this program to facilitate Colombia's own Plan Patriota—a military frontal assault aimed at the FARC. Washington has been far softer, even conciliatory, to another group that also it has labeled as terrorists—the right-wing AUC paramilitaries, which share many U.S. strategic objectives in Colombia, such as the annihilation of left-wing guerrillas. This element has weakened the credibility of any antiterrorist campaign, a point that will be developed later in this chapter and in the subsequent one.

In terms of operationalizing the "war against terror" in Colombia, the first step occurred between October 2002 and January 2003, with the arrival of a U.S. counterinsurgency team in Colombia. It provided training to some 4,000 Colombian troops and also presented ten helicopters and other materiel to the Colombian army in efforts to fight the FARC.[9] An important and related goal was to train the Colombian army within the space of a few years, in order to facilitate a relatively quick retraction of American forces. Between 1999 and 2004, 32,458 Colombians received U.S. military/police training, and by 2003 Colombia was the recipient of more U.S. training than any other country.[10] Since official U.S. antiterrorism aid was implemented in Colombia during 2002, Colombia's Urban Antiterrorism Special Forces have been fortified. Moreover, Washington has provided training for Colombians through the U.S. Counter Terrorism Fellowship Program, for which Colombia received a grant of some $300,000 in fiscal year 2005.[11] The "antiterror" discourse also has provided the United States with a justification to use part of PC to militarize Colombian petroleum installations in an attempt to secure oil for American consumption.

Beyond the issue of terror, a major strategic problem for Colombia historically has been the lack of a government presence in much of the country. In an attempt to address this huge void, President Uribe created a Peasant Soldiers brigade comprising some 15,000 recruits, who received three months of military training.[12] The president had envisaged the training of up to 100,000 peasant soldiers by the end of 2006.[13] But this has not been complemented by any social welfare program in these regions. Thus, locals have seen only the coercive apparatus of the state, without the benefits of schools, hospitals, and so on. Finally, the related aspect of the privatization of warfare is worth noting. It is important to emphasize that a significant portion of training received in Colombia has been conducted by U.S. private contractors, a point to be developed in the next chapter.

Asymmetry

The Colombian case demonstrates the concept of asymmetry in a number of ways. It is evident with respect to issues regarding organization, strategy and tactics, weaponry, timelines, and epistemology. The FARC has been organized into dispersed fronts of 150–200 well-armed and well-funded soldiers, which means that it is not vulnerable to a centralized attack.[14] As we shall see, PC's extensive use of surveillance has meant that the FARC frequently has traveled in units of about 25 or less soldiers to avoid clear detection. This distinguishes them not only from national armies but also from groups such as Peru's Sendero Luminoso. That rebel group was built around the cult of leadership associated with Abimael Guzmán, whose capture meant Sendero's essential dissolution. Thus, one aspect of the RMA as manifested in Colombia is the network organization of the FARC and other Colombian belligerents, and the related reliance on "codes" that connect dispersed, clandestine units that are not vulnerable to a centralized attack.

The FARC and ELN have deployed different weapons than the Colombian and U.S. governments. They have relied on assault rifles (such as AK-47s), grenades, cylinder bombs, and land mines. By 2004 there were an estimated 100,000 land mines deployed defensively by guerillas, which were responsible for about one-third of the casualties among the armed forces.[15] The FARC has resorted to horse bombs and bike bombs. Further, the downing of various Colombian and private military aircraft shows that the rebels are likely to possess shoulder-launched surface-to-air antiaircraft missiles. All this, of course, stands in contrast to the United States' high-tech weaponry, sophisticated satellite surveillance, and so on.

In terms of tactics, rather than direct confrontation, the FARC and ELN have utilized classic guerrilla strategy that employs hit-and-run attacks, as well as occasional acts of terrorism. There has also been an emphasis on nonlinear combat, with respect to both space and time. With regard to temporal strategy, for example, the FARC retracted from military confrontation during much of 2004, leaving observers to wonder whether the group was severely weakened by PC, or whether the rebels were simply playing possum. The FARC's remarkable resurgence during early 2005 and through 2007, involving major surprise attacks against the Colombian military in a variety of locations around the country, suggests their appreciation of nonlinear warfare. This was expressed through an interrupted and unpredictable sequence of combat, and through a choice of dispersed armed confrontations such that the next attack seemed as if it could occur anywhere and anytime and with no obvious linear pattern.

Another asymmetric element includes the different timelines assumed by the U.S. and Colombian governments on the one hand and the Colombian rebels on the other. The United States officially was scheduled to terminate PC effective late 2006, when it had hoped that the FARC would be weakened into assuming negotiations favorable to the Colombian government. The FARC, however, has viewed Colombia to be in a state of continuous, historical warfare, with PC representing an important but relatively brief phase that the revolutionaries were prepared to wait out. The element of strategic time, then, represents a crucial distinction between the competing sides.

A final element of asymmetry concerns the epistemological distinctions between the United States and the Colombian belligerents. The United States has formulated a strategy that blends modern elements, such as the fortification of the Colombian state and military to defend national territory, with some postmodern elements that will be discussed shortly. By contrast, there exists a distinct patchwork of epistemological elements in Colombia, ranging from those associated with premodernity to others linked to postmodernity. For example, Colombian belligerents and much of the population have continued to think of political space in a nonmodern fashion that emphasizes decentralized, locally controlled, fragmented parcels of land. The guerrillas and paramilitaries, to a significant extent represent premodern, agrarian interests (both legitimate and illicit crops). As well it was observed earlier that they have attempted to exact a feudal-style of tribute in exchange for the provision of security. Beyond these premodern or nonmodern elements, they have embraced the postmodern economy of transnational crime While these features and others are dealt with more fully elsewhere,[16] suffice it to say here that it would be strategically perilous to avoid consideration of the epistemological dimensions of asymmetry. It is essential to come to grips with the epistemological foundations that guide the actions of various actors, and to appreciate the strategic implications of the distinctions between them.

The Colombian army has been restructured by the Americans to deal with asymmetric aspects of warfare. The challenge was how to refashion the Colombian military to fight a highly successful and dispersed guerrilla group. Colombian forces needed to be quicker, more mobile, equipped with better intelligence, more capable of moving into difficult terrain such as high mountains and also rivers, and able to fight at all times including nighttime.[17] The result of this restructuring was expressed most clearly in Colombia's Plan Patriota, which deployed approximately 17,000 troops to debilitate the FARC. The first phase occurred during June to December 2003 and was concentrated in the capital of Bogotá and in the department of Cundinamarca. The second and more difficult segment, which began in February 2004 and continued through 2006, focused on the FARC's heartland of Caquetá, Meta, and Guaviare. A third stage was to be launched in Antioquia by the end of 2005, but was delayed due to complications associated with the FARC's resurgence. In December 2006, after a series of devastating attacks by the FARC that demonstrated the group's resilient military capacity, President Uribe declared an end to Plan Patriota and announced Plan Victoria—similar in style to Plan Patriota but designed to defeat the FARC, according to the president.[18] Both plans involved assistance from the U.S. Southern Command, the U.S. Department of State, and the U.S. Department of Defense.

The United States has provided a wide assortment of training and materiel to support the Colombian military in its various endeavors. This includes fighter aircraft such as the C-26 and the AC-47, in addition to many others used in coca crop spraying, such as the OV-10, the T-65, and the AT-802. Also included in the package were specialized helicopters such as the Huey II and UH-60 Blackhawks (some 65 helicopters had been supplied by 2003).[19] The United States also supplied interdiction boats for riverine combat. This is key, since the illicit arms and drugs on which the FARC have depended are often transported through Colombia's extremely lively river routes. Importantly, PC supplied significant logistics and communications equipment to Colombian forces, while the United States operated an elaborate real-time satellite surveillance system from which it provided intelligence to Colombians. Other assistance included aerial and ground radar systems and infrared devices,[20] in addition to night-vision goggles.[21] In terms of operational restructuring, U.S. assistance from PC offered equipment and training for the formation of Colombia's first rapid-action brigade, the Fuerza Despliegue Rápida (FUDRA), three high-mountain battalions, and an assortment of crews specializing in riverine battle.[22] The Colombian government claimed that by February 2006 Plan Patriota had resulted in the killing or severe injury of 2,518 guerrillas.[23]

Plan Patriota was designed to address asymmetry by permitting the Colombian army to be more informed, more mobile, quicker, and generally more able to fight on almost any terrain at any time. But perhaps the most important weapon on the side of the U.S. and Colombian governments to cope with asymmetry in their fight against leftist rebels has been the proliferation of the paramilitaries. Clearly, the paramilitaries have shared strategic objectives with the U.S. and Colombian militaries to the extent that they wish to eliminate the guerrilla Left. The AUC has been able to fight the FARC and ELN on an equal footing—it simply pits a left-wing guerrilla force against a ferocious right-wing guerrilla army. The paramilitaries have a record of fighting "dirty," and are not bound to the international criticisms and sanctions that could be directed against state armies if they behaved in a similar fashion.

The accentuated paramilitarization of Colombia is at minimum a striking coincidence that is highly supportive of the key objectives of PC, such as the capacity to engage effectively in asymmetric warfare. For example, there have been noteworthy operations whereby the Colombian military moves into certain zones or cities and pushes out the FARC or ELN, with the area subsequently falling under the control of paramilitary forces. This has occurred since 2002 in Medellín, Colombia's second most important city, and since 2000 in Barrancabermeja, the oil capital of the country. In fact, one of the most remarkable strategic developments during 2004–2007 has been the advancement of paramilitary control of key urban areas, a shift that has been starkly obvious to numerous observers.[24] It was revealed in March 2007 that the CIA had evidence linking Colombia's army chief and close ally to President Uribe, General Mario Montoya, to a paramilitary group involved in assuming control over parts of Medellín following the aforementioned 2002 military assault of guerrillas in that city. The CIA document suggests that the paramilitaries, the Colombian police, and the army jointly planned the 2002 operation in Medellín.[25] Here, then, we observe a nexus between the Colombian government, the military, and a paramilitary group linked to terrorism, countless human rights abuses, and drug trafficking. Officials in Washington and the U.S. embassy in Colombia would have to be extremely naïve not to have noticed this nexus all along, as it certainly addressed the asymmetric component of Colombian warfare.

Besides this, both the U.S. and Colombian governments have embarked on an interesting discourse whereby FARC members are viewed as "bad" terrorists, whereas members of the right-wing AUC are portrayed as relatively "good" terrorists capable of social rehabilitation through the demobilization process. In a

U.S. State Department bulletin in 2004, for example, Washington highlighted its cooperative efforts with the Colombian military, which had begun "to strike hard at the FARC's leadership ranks," while praising "peace talks and a pilot demobilization program with large elements of the AUC."[26] We shall return to an analytical consideration of the demobilization process in Chapter 4, but suffice it to say here that the paramilitaries represent the state's antidote to asymmetric warfare in Colombia.

Complexity

Complexity emphasizes the interrelatedness of human actions and suggests an analysis that avoids viewing its subject in isolation. In Chapter 1 we observed that the RMA denotes a system of systems approach, wherein a wide variety of themes are complexly interrelated. This implies an interdisciplinary perspective to strategic affairs, one that includes elements of politics, economics, ecology, geography, literature and art, the sciences, and so on. Further, a key aspect of strategic complexity concerns globalization. It can be viewed as a concert of three strategic components, including time, space, and movement. With great foresight, just prior to the era of globalization, Virilio observed that "[r]evolution is movement. . . Politics is only a gear-shift, and revolution only its overdrive: war as 'continuation of politics by other means' would instead be a police pursuit at greater speed, with other vehicles."[27] Complexity, then, can be related to what is moved, the ease of movement, volume of movement, rapidity of movement, distance and range of movement, as well as (in)visibility of movement. The RMA is situated within the context of globalization and the profound revolution of movement on which it is based.

With regard to Colombia, the theme of movement is paramount. Movement has helped define eras of politics and warfare in the country, from the slow-mo[ving,] obstacle-filled, and introverted world that lasted into the beginning of the twentieth century, to the period of greater commercial and ideational movement from the 1920s through the 1960s, and to a globalized world that commenced in the 1970s and 1980s. It was during the epoch of globalization that transnational crime blossomed and began to feature the voluminous but veiled movement of drugs, arms, and money over great distances and with increasing acceleration. Its illegal nature, which initially imposed an impediment to ease of movement, in turn induced a degree of profitability that identified relatively facile clandestine routes. It was this revolution in movement, time, and space that so changed the nature of the Colombian imbroglio since the 1980s. It has enriched and empowered competing belligerents to the point that war may no longer be winnable.

There are a plethora of other features associated with the complexity of the Colombian case. Among them is the task of winning a war in a country that never had a workable nation-state or a Leviathan. This means that state building must occur alongside intense warfare, which is further complicated in an era of neoliberalism that promotes government cutbacks rather than political construction. The diversity of contestants, the pronounced political and geographical fragmentation, and the difficulty in identifying political good guys also contribute to the complexity of Colombia's imbroglio. Another classic manifestation of complexity concerns the enormous extent of aerial fumigation of coca crops in the country. Complex results of this include an important environmental impact, the balloon effect, which pushes the trade outward, and the political push of *cocaleros* into the strategic embrace of the FARC—points that will be explored in the next chapter.

Surveillance and Intelligence

It was noted in Chapter 1 that the notions of intelligence and spying have been celebrated strategic concepts that dotted the classics, such as Sun Tzu's *Art of War,* in which he observed that only the best and brightest should be bestowed with the crucial task of spying. Further, Bentham's *Panopticon* signaled a new instrument of power over the human mind and body—the use of surveillance. In this section, we shall address two broad questions related to surveillance and intelligence. First, what are the successes and limits of ultra-surveillance and real-time transmission of information as reflected in PC? Second, beyond technological gadgetry, how have human intelligence services been employed in Colombia, and to what effect?

Let us begin with real-time satellite and radar surveillance. In 1999, a year after the FARC received its *zona,* real-time aerial surveillance began to be used extensively in the Colombian conflict. This intensified under PC[28] and involved the use of Plataforma aircraft that come equipped with heat sensors capable of detecting human activity even at night.[29] PC's debut in 2000 also included $31 million from the U.S. Department of Defense for aerial antinarcotics intelligence, land radar systems, command and control systems for radar, and translations of intelligence analyses; $17.4 million from the U.S. Department of State for aerial antinarcotics intelligence, logistical support, and communication between operatives; and $5 million divulged to the Colombian National Police for better communication between the police and armed forces and for improved logistical support. Night-vision equipment have also been included in the PC package, because it is estimated that about 80 percent of key FARC operations have occurred at night.[30]

Further, PC included radar sites located in strategic regions, especially in the south of Colombia, which has been the heartland of the FARC's support. Among these are three ground-based radar systems in the Amazon basin near Leticia, a base at Marandúa, Vichada, and two others at San José Guaviare and Tres Esquinas. A couple of other radar sites located in Colombia, which are part of the U.S. Air Force Caribbean Basin Radar Network, also have been utilized for PC, including bases at Riohacha and the island of San Andrés (located off the coast of Nicaragua). Besides the ground-based systems, the United States provided forward-looking infrared radar systems for Colombian aircraft.[31] Moreover, it is worth emphasizing that a number of private U.S. military contractors have benefited from supplying intelligence equipment for these projects, such as Lockheed Martin, Dyncorp, Arinc, TRW, and Integrated AeroSystems.[32]

PC also entailed the creation of a number of forward operation locations (FOLs) in Colombia and in the surrounding regions. These bases provide logistical facilities that store weapons and also serve as stations for aircraft and personnel. FOLs located outside of the country have included the Eloy Alfaro Airport in Manta, Ecuador (Ecuador's president Rafael Correa threatened in 2007 to close this site, though it cost the United States $61.3 million. It was created after the U.S. military was expelled from the Howard Air Force Base in Panama),[33] the Hato International Airport in Curacao, the Reina Beatriz Airport in Aruba (developed at a cost of $10.3 million), and the Comalpa International Airport in El Salvador. Subsequent to these initial FOLs, another one was constructed in Saravena, Arauca, to support U.S. oil interests, and additional FOLs were requested for fiscal year 2006 in various conflict zones throughout Colombia.[34]

Even countries that are not typically associated with supporting U.S. strategic objectives have been involved in the surveillance component of PC. For example, Brazil completed a controversial $1.5 billion satellite surveillance system to protect against the inflow of drugs, arms, and subversives from Colombia in the wake of PC's implementation. Much of the technology for this massive project was supplied by the U.S. military contractor Raytheon.[35] Further, in 2004, Brazil agreed to a U.S. request to shoot down aircraft suspected of drug trafficking, triggering widespread debate throughout Brazil and the rest of South America due to a disastrous episode in Peru in which a planeload of missionaries mistaken for drug traffickers were shot down in April 2001.[36]

The use of real-time surveillance has not been without controversy in Colombia, especially after an incident in 1998. That is when U.S. military contractor Airscan, hired by Occidental Petroleum and Ecopetrol, provided intelligence to a Colombian fighter aircraft regarding the presence of guerrilla activity in oil-rich Santo Domingo, Arauca. The bombing raid mistakenly targeted civilians, killing 18 innocent Colombians and wounding scores of others. For five years the U.S. contractors and the Colombian air force blamed each other for the disaster, until a video was released in 2003 that showed two U.S. contractors shouting at Colombian helicopter pilots to stop firing at civilians. That watershed triggered the resignation of the head of the Colombian air force,[37] and raised doubts among U.S. leaders about providing the Colombian military unlimited access to real-time surveillance information. Credible fears of infiltration of the Colombian armed forces by the guerrillas or paramilitaries, in tandem with the Colombian military's poor record with respect to human rights abuses, compounded American concerns about sharing real-time intelligence with Colombian authorities.[38] In this context, the Clinton administration, with Presidential Decision Directive PDD 73, limited the sharing of U.S. intelligence with Colombia except in cases dealing directly with drug trafficking. Following 9/11, when the Bush administration used the War on Terror to justify overtly linking PC to the war against Colombian leftist guerrillas, the United States revised that law to share more, but not all, real-time information with the Colombian military in order to weaken the FARC and the ELN.[39]

All this led to a public campaign by the Colombians to request complete access to U.S. real-time intelligence, in addition to loud complaints regarding the provision of delayed information that limited the capacity of the Colombian military.[40] This triggered U.S. officials publicly to dance around the issue, exemplified when

U.S. ambassador Anne Patterson told the Colombian media that "[s]ome people think we have a secret satellite that has everything. That is crazy . . . (it is) very difficult for us to collect intelligence, as well."[41] Insiders suggest that, especially in the wake of the Bush amendment to the Clinton-era law noted above, Colombia indeed has received greater amounts of real-time intelligence, but has remained quiet about it given the assortment of associated controversies.[42]

Under ideal conditions, satellite surveillance can spot human activity on the ground at a resolution of between one and nine meters. But this does not mean it can see everything of strategic significance, all the time. For example, dense cloud and fog limit such surveillance, as does thick jungle foliage.[43] Given the prevalence of both clouds and jungle in Colombia, the impact of this limit should not be underestimated. Under such circumstances, guerrillas are naturally drawn to cloudy areas and to leafy jungle terrain. Further, even under good conditions, satellite surveillance cannot necessarily distinguish guerrillas from other people, except for telltale signs. Prior to such surveillance, the FARC was able to travel in large units sometimes numbering over 100 troops. But groupings of that size have rendered them conspicuous to satellite surveillance, forcing them to travel in less obvious bands of 25 or so. Moreover, while aerial surveillance can detect human movement in wide, open spaces and in the countryside, it cannot peer into buildings. Thus, there has been a propensity for guerrillas to move more extensively into clandestine urban terrain, with universities evolving into favorite targets.[44] Also, there is not the logistical capacity to view all Colombian ground space all the time. Limited human resources mean that surveillance is discretionary, in the sense that one has to know where and when to look. Finally, human error associated with surveillance data can lead to disastrous consequences, as demonstrated by the 1998 Santo Domingo tragedy.

* * *

The effects of real-time aerial surveillance are mixed from the perspective of the U.S. and Colombian governments. On the positive side of the ledger, it has fortified the Colombian military and thus has provided Washington and Bogotá with an edge in their battle with leftist guerrillas. For example, it has made it more difficult—though still very possible, as we shall see below—for the FARC to attack Colombian military bases, as this previously had been accomplished by large numbers of FARC troops that potentially can be spotted with surveillance technology. Importantly, real-time aerial surveillance, together with night-vision technology, has facilitated the deployment of Colombian rapid-action forces. This is part of the reason behind an increasing level of Colombian military attacks on rebel forces, rising from 477 in 2002 to 1,748 in 2004.[45] Also, this sort of intelligence has provided the U.S. and Colombian governments a greater, though still limited, capacity to spot flow of illicit drugs, arms, and money that benefit the guerrillas. The U.S. and Colombian militaries have also been strengthened by the potential to intercept phone and fax messages, e-mail messages, and so on. Further, private security companies hired to protect major corporate interests in Colombia, especially in the extractive sector, have used electronic devices such as real-time cameras, infrared surveillance, and movement detectors. The information thus gathered not only has served to protect them from guerrilla assaults, but also in many cases has been turned over to Colombian and U.S. authorities.[46] Despite the positive impact of surveillance for the U.S. and Colombian militaries, there is still room for improvement. A series of disturbing and destructive cases of friendly fire on the part of the Colombian military suggest that the degree of intelligence still needs to be improved considerably. One such case involved a military ambush on a Colombian police unit that killed seven policemen and four prisoners in March 2004.

Beyond the high-tech wizardry of aerial and other electronic surveillance, it is important to emphasize that human intelligence is at least as important in the Colombian case. In fact, it is the mixture of high-tech and human surveillance that is key. There are a number of dimensions of human intelligence associated with PC. One of these is elite spying units, which have mixed U.S. and Colombian personnel. For example, Colombian agents had U.S. assistance in tracking the FARC's Ricardo Palmera, or Trinidad, during Christmastime 2003–2004. Palmera, one of the FARC's major financial leaders, was apprehended in neighboring Ecuador.[47] The U.S. and Colombian spying units have collaborated in the capture of other top-level FARC leaders.

An intriguing component of President Uribe's extensive reliance on human intelligence is Plan Meteoro, whereby certain highway travelers have been equipped with panic buttons and other communication devices linked to satellites. This has entailed the transmission of information regarding guerrilla activity on the country's highways, which have been notoriously perilous due to rebel control of large swaths of Colombian territory. The purpose was to render the roads safer for public and commercial use, especially during holidays. It was essentially designed to permit freer movement of people and goods and, at least in temporary instances, to reclaim space and transit ways for the Colombian population at large.

More troublesome is a novel intelligence system implemented during President Uribe's tenure involving the creation of what he claimed was 1.5 million informants throughout the country who could phone a toll-free number to report suspicious guerrilla activity. Imagine how much one knows about one's neighbors or fellow coworkers—Uribe's plan was brilliant to the extent that it attempted to harness accurate local gossip and to transform it into priceless strategic intelligence. But much of this information appears to have been inaccurate. For example, approximately 3,600 people were arrested during July 2003 to July 2004, but 80 percent of them were released after lengthy detentions.[48] This tactic was based on the pernicious premise that one is guilty until proven innocent. Further, many strategic experts indicate that the number of 1.5 million is greatly exaggerated and that there does not exist in Colombia a logistical capacity to deal with that many informants.[49] However, this plan represents the first time the Colombian government has attempted such an extensive human intelligence operation.

Further, suspicions have been raised by keen observers that many such informers were actually members of paramilitary organizations. They likely have divulged such data to suit their own agenda of consolidating power in various regions of the country.[50] This information, then, may have assisted the AUC with its objective of conquering strategic territory from the FARC and the ELN—particularly resource-rich areas or land suited to coca cultivation. It may also be related to the Colombian military's incursions into urban areas controlled by leftist guerrillas, such as the cases of Medellín and Barrancabermeja mentioned earlier.

There are legitimate complaints that enhanced human intelligence associated with PC has resulted in significant human rights abuses. Linked to this kind of surveillance, for example, are *captivas masivas* (massive captures), whereby entire communities are detained because some of its members are suspected of being guerrillas or of supporting them. Such detentions typically have lasted weeks or months. A number of experts have criticized the strategic implications of this process. Chris Patten, a commissioner for foreign relations in the European Union, indicated that this approach was tried in northern Ireland in an attempt to weaken the Irish Republican Army. But it failed miserably as it provoked public hostility toward government forces. A local expert has argued that "if the intelligence is not sufficiently precise, there is a tendency to charge people who are innocent," and thus turn the population against the state. In Colombia, between October 2002 and January 2004, 850 people are on record for being detained in such circumstances, with 414 released and 53 cases tied up in court.[51] Further, President Uribe's antiterror statute, passed by the Colombian Congress on 11 December 2003, permitted members of the military to conduct searches, intercept communications, and detain individuals without a judicial order.[52] Thus, surveillance systems clearly possess the capacity to abuse civil liberties and human rights.

Surveillance, Guerrillas, and Social Forces

The FARC compensated for the extensive use of surveillance by the U.S. and Colombian militaries by exploiting the limits of surveillance and also by engaging in its own adventures in surveillance. The elements of fog and jungle foliage were already mentioned, as was the enormous logistical limit to being able to see everything all the time. In this evolving game of cat and mouse, guerrillas gravitated toward notoriously cloudy areas and heavy jungles to escape satellite detection. Importantly, satellite surveillance is most effective in the countryside and can do little to spot guerrilla activity in urban areas. Hence, the FARC responded by a push into Colombia's major cities. Evidence of this includes the FARC's attack of the presidential palace during Uribe's inauguration on 24 October 2002, the placement of a briefcase bomb in an annex of Hotel Tequendama on 13 December 2002, and five fire bombs placed at Bogotá's Transmillenio on 6 January 2003. Most important was the FARC's notorious attack in February 2003 on the Bogotá's posh El Nogal Social Club. It demonstrated that the rebels could penetrate even the toughest private security, and that the capital city's upper class was not immune to war. These sorts of attacks eroded popular support for the FARC, who were portrayed by the government as urban terrorists. Although there have been occasional small-scale urban attacks launched by the rebels since then, more recently, their urban endeavors have shifted to the less conspicuous task of university recruitment.

Perhaps most importantly, the FARC has adopted a policy of occasional strategic retreat, hiding where it can from government surveillance and thus conserving its forces until an anticipated wavering of U.S. pressure. The FARC emerged from retreat in early February 2005, with remarkably successful attacks on Colombian military bases and towns in Vista Hermosa (Meta), Santa Ana (Putumayo), Iscuande (Nariño), and Mutatá (Urabá). These ambushes left 45 Colombian soldiers dead in nine days. Further, during March and April 2005, the FARC clearly demonstrated that its power had not cascaded substantially when rebels successfully attacked the towns of Toribio and Tacueyó, in Cauca, embarrassing Colombian military forces who were unable to repel the assault. Similar attacks continued throughout 2006 and 2007. Strategically, the rebels' reemergence was designed to demonstrate the "failure of the Urbista war," which the guerrillas claim has not diminished the military capacity of the FARC.[53]

Besides the FARC's ability to compensate militarily for increased intelligence associated with PC, there are other obvious limits to government-sponsored surveillance. First, the FARC itself is thought to have made substantial investments in both communications and intelligence.[54] Illustrative of this was the FARC's interception of supposedly clandestine electronic communication between the Colombian army and the intelligence service Departamento Administrativo de Seguridad (DAS). This resulted in a rebel attack at Hacarí in April 2006 that killed seven military members and ten DAS employees.[55] Given the substantial coffers of these guerrillas, they are no doubt capable of purchasing sophisticated technology and of concocting generous bribes in the realm of human intelligence. This may counteract to some extent the intelligence instruments employed under PC.

The ELN, clearly the junior leftist insurgent group in Colombia, has been decidedly less successful than the FARC with regard to compensating for the use of surveillance and other aspects of PC. Its troops are estimated to have dwindled from an estimated high of about 5,000 in the year 2000 to less than half by 2007. This is not so much a result of surveillance employed on the part of the U.S. and Colombian militaries, but has more to do with encroachments on the ELN by both the FARC and the AUC. While the ELN has aligned with the FARC since about 2003 in certain areas, it has lost its power to the FARC especially in its key strategic zones of oil-rich Arauca and Norte de Santander. For example, the FARC in June 2006 announced an offensive against the ELN in Arauca. Most importantly, the ELN has been the victim of unrelenting attacks by the AUC. Both the FARC and the AUC have employed their own surveillance tactics against the ELN.

Surveillance, the Paramilitaries, and Social Forces

There is little evidence to suggest that U.S. or Colombian intelligence forces have been directed in any substantial way against their strategic ally, the AUC. This is especially the case since 2004, when the AUC entered negotiations with the Colombian government regarding a demobilization process designed to incorporate AUC members into "legitimate" society. There are a number of cases that appear to demonstrate collusion between the paramilitaries and government intelligence services. A case in point concerns the firing in October 2005 of two DAS directors for selling intelligence information to the AUC.[56] Further, President

Uribe has admitted that former Venezuelan soldiers plotted the overthrow of President Hugo Chávez at the Bogotá headquarters of DAS in an operation that likely involved paramilitary forces.[57] Moreover, major labor unions broke off talks with the government in April 2006 after it was revealed that DAS authorities provided intelligence information in the form of a "blacklist" of union leaders to AUC members. The director of DAS, Jorge Norguera, was arrested in February 2007 for allegedly providing information regarding union leaders to paramilitary death squads.[58] Thus, there appears to have been a cooperative supply of intelligence among the AUC and government forces. We shall return to the issue of collusion between the government and the AUC in the next chapter when we address the apparent unraveling of the demobilization process in late 2006 and into 2007.

Surveillance and Civil Society: Labor and the Displaced

While the next chapter will provide a wider array of statistics and analyses regarding human rights abuses directed against labor, a thumbnail sketch is provided here as a backdrop to the issue of surveillance. While the number of assassinations of union members was down from previous levels during 2003–2005, the number has risen for selective assassinations, detentions, threats, and hostage taking.[59] For example, homicides of union members decreased from 198 in 2001 to 70 in 2005. But threats against union members rose from 234 to 260 and detentions grew from 8 to 56, while hostage taking of union members increased from 13 to 32 during the same period.[60]

Key members of labor organizations such as the oil union USO, and of umbrella unions such as the Confederación General de Trabajadores Democráticos (CGTD) and the Central Unitaria de Trabajadores de Colombia (CUT), regard the paramilitaries as their chief enemy. They view the AUC as the perpetrators of most of the assassinations of and other human rights abuses against organized labor. They also see the AUC as a chief provider of intelligence to government forces, which routinely detain union members on suspicion of subversion or of supporting leftist guerrillas.[61] Earlier we observed allegations regarding a blacklist of union leaders provided by DAS to the AUC in April 2006. Thus, there are credible assertions by many labor leaders of intelligence linkages between the state, the military, and DAS that are designed to suppress the political activity of unions. These leaders argue that much of the intelligence information is false, and has been part of an attempt to intimidate those who oppose the privatization of state corporations as well as the promotion of transnational corporate interests that are viewed as antithetical to those of labor.[62] We shall return to this general issue in the next chapter, which addresses the relation between privatization and PC.

The effects of surveillance on the displaced population are indirect but crucial. For example, one disturbing component of the increased intelligence and military force associated with PC has been the alarming increase of *bloqueos*, whereby local populations are forced by belligerents to remain in specific locations for strategic purposes.[63] Thus, increased surveillance associated with PC appears to have limited the range of movement by leftist guerrilla groups. Rather than the previous unifocus on conquering new territory, the FARC and ELN have adopted a strategy of attempting to consolidate their existing power. In addition to leftist rebels displacing populations through territorial invasions, then, there has been a parallel tendency toward *bloqueos,* whereby guerrillas attempt to control more stringently populations and territory already under their domain. *Bloqueos* also entail an effort to halt arms flows antithetical to rebel interests, to prevent the outflow of strategic information, and so on. Thus, the development of *bloqueos* has depressed displacement figures to some extent, but suffering on the part of innocent civilians has not been reduced. *Bloqueos*, induced in part by increasing surveillance, are simply the equally pernicious flip side of displacement. Thus, one cannot view decreasing levels of displacement as a necessarily positive sign when they are complemented by the increasing phenomenon of *bloqueos.* Related statistics and a more rigorous analysis of displacement will appear in the next chapter.

Surveillance, Strategy, and Political Economy

Surveillance has served as an instrument of power in the realm of political economy, as exemplified by the role of IMF in Colombia. With the country's debt standing at about $37.21 billion (US) in 2006, a figure that has remained steady since 1998, the task of the IMF has been to survey the Colombian economy for the broad purpose of debt restructuring. Related neoliberal policies have been enacted, including privatization, liberalized trade, enticements for foreign investors, and the imposition of limits to social welfare spending. While Colombia is in dire need of a state that works—one that fosters basic education, health care, and security throughout the country—IMF policies are aimed principally at attracting transnational investment. These two objectives are not necessarily contradictory, but in the Colombian case corporate interests have been served while the tasks of state building and the reduction of severe inequity and poverty have been ignored. On the positive side of the ledger, such policies have contributed to Colombia's GDP growth rate rising from 1.5 percent in 2001 to 5.1 percent in 2005. In 2006 the country hit its highest growth rate in 28 years at a booming 6.8 percent. Investor confidence also has risen in Colombia, and this is especially true in the extractive sector within the context of increasing regulation and nationalization in countries such as Bolivia and Venezuela. Foreign investment in Colombia rose 33.1 percent in 2006 to $2.675 billion (US).

But the benefits associated with macroeconomic growth and rising foreign investment clearly have not trickled down to the country's majority population. Even with the 6.8 percent GDP growth rate in 2006, unemployment rose to 11.8 percent from 10.4 percent in 2005, and subemployment rose from 31.2 percent in 2005 to 35.7 percent in 2006. Thus, the benefits of the economic boom have been concentrated at the top.[64] Further, the already high levels of poverty and inequity in Colombia have worsened. These have been exacerbated by IMF policies that urge welfare cuts and also attempt to shrink an astonishingly anemic state. Poverty affects about 65 percent of Colombians, as it has for years. Poverty rates in rural areas—the breeding grounds for guerrilla support—hovers at about 68 percent.

These rates would be much higher if it were not for the remittances sent home by the 10 percent of the Colombia's population who have fled to other countries. Remittances were worth $55 billion (US) in 2005.[65] Colombia's gini coefficients reveal rising inequity over the last three decades. Ginis were 0.48 in 1978, 0.55 in 1991, 0.59 in 1999, and 0.58 in both 2004 and 2005. The United Nations Program for Development (UNPD) measured inequity in Colombia as among the worst in the world, with a rank of 119 out of 127 countries surveyed in 2004. In 2005, the UNPD estimated that 22.6 percent of Colombia's population lives on less than $2 a day, with another 8 percent living on less than $1 a day. This is against the backdrop of the poorest 10 percent of the population receiving 0.8 percent of the country's wealth, while the top 10 percent received 46.5 percent.[66]

Given such a context, IMF policies, such as a 2004 agreement that limited the state's deficit to 2.8 percent of its GDP, are strategically counterproductive to the extreme.[67] So, too, are all-too-frequent pronouncements by the agency that express its preoccupation regarding Colombian debt levels, with no mention of the importance of state building in this war-ravaged country.[68] This is not to advocate mounting levels of debt, but only to emphasize that debt restructuring needs to be placed within a wider social and strategic context. In short, the IMF and the interests it represents have promoted policies that appear to benefit corporate interests, but have not addressed the profound poverty and escalating inequity in Colombia. With its absence in much of the nation's territory, it is elementary that Colombia badly needs a bigger state, not a smaller one.

Financial surveillance can be highly positive and yield basic transparency that can benefit social development. Surveillance, for example, can be harnessed for particular projects such as tax collection. Colombia has only recently begun, in any serious manner, to collect taxes from the wealthy and has imposed new penalties for tax evaders. A war tax also has been introduced. All this meant a record level of taxes collected since 2004. The subsequent issue, of course, is how these taxes are spent. The state cannot successfully rely solely on repression but must provide basic services throughout the country to win the hearts and minds of the population—a classic strategic formula that is strangely absent in PC. The issue of surveillance and international finance will be revisited in the next chapter, which addresses privatization and warfare.

Surveillance and Global Civil Society: Why Colombia Remains Ignored

Why do the media, global NGOs, and international social movements pay so much attention to certain parts of Latin America but ignore others? Given the huge U.S. military investment in Colombia, in tandem with the enormity of human rights abuses there, one might expect far greater global interest in the country than there has been. There are many reasons for this. To begin, Colombian NGOs are far less developed and organized than their Mexican counterparts, largely due to the dire threats they face from the country's assortment of belligerents. That is, the climate of intense fear and violence has stymied the development of a free-speaking civil society that can dedicate itself to establishing global connections. Besides the general weakness of Colombian civil society, the country's array of criminally violent guerrilla groups and paramilitary forces has not had much appeal for global NGOs and social movements. The global audience, it appears, is more attracted to clever and nonviolent revolutionaries, such as Mexico's EZLN, and to the radical Left that comes to power through the ballot box, such as Hugo Chávez, Evo Morales, and Rafael Correa. Further, through the use of neomercenaries and paramilitary forces, the United States has reduced its own troop commitment, thereby minimizing American scrutiny. In the next chapter, we shall develop these points and introduce others related to the theme of surveillance and social forces.

Conclusion

> But they never found the sea.[69]
>
> *Gabriel García Márquez,* One Hundred Years of Solitude

PC reflects the RMA in a number of ways. One of these concerns terror and the discourse surrounding it. We observed that it has been a handy instrument of power for the U.S. and Colombian governments vis-à-vis leftist rebels in the country. We shall see in the second half of the book that a similar pattern is discernable in the Mexican case. Another important and wide-reaching element of the RMA is asymmetry. It entails distinctions among actors related to organization, strategy, weaponry, conceptions of time, and epistemology. With regard to Colombia the emphasis has been on asymmetric struggle between the state and insurgent forces. In the Mexican case, as we shall see, asymmetry is perhaps more pronounced in relation to the struggle launched by resistance movements and critical NGOs. Beyond asymmetry, Colombia demonstrates the theme of complexity. Much of this is related to a revolution in movement. Another dimension of complexity worth emphasizing in both Colombia and Mexico is the assorted causes and effects of narcotrafficking, a theme revisited in the chapters ahead. Finally, surveillance represents a crucial instrument of power that is clearly linked to the RMA. We considered both technological and human aspects of surveillance. Colombian civil society has not been as adept at employing strategic surveillance as its Mexican counterpart, for reasons we shall explore in Chapter 7.

What is important to emphasize here is that it is the synergetic influence of various features of the RMA that helps explain strategic change. That is, while each of the components of the RMA is important in its own right, the interaction between them has had an exponential effect. For example, the discourse on terror has permitted the U.S. and Colombian governments to be entirely clear that PC has been aimed at routing the FARC. The restructuring of the Colombian military to fight asymmetric warfare against the FARC, complete with all-terrain and rapid-action forces, has been guided by an array of new ultra-surveillance systems. These same surveillance systems have been used to protect U.S. interests in Colombia's petroleum reserves and to bolster neoliberal discipline that renders

Colombia to be a draw for foreign investment in the extractive sector. The discourse on terror, asymmetric warfare, and surveillance are part of a system of systems that has been designed to diminish the power of the FARC, to promote U.S. economic interests in the extractive sector, and to place U.S. forces on a strong military footing in a region of the world that has tilted to the Left as have Colombia's two neighbors—Ecuador and Venezuela.

What has all this meant for social forces? PC has affected the labor movement in a number of ways. PC has accompanied neoliberal restructuring arrangements that are generally antithetical to the interests of union members, a point that is further explored in the next chapter. Components of the plan, especially surveillance, have been used as instruments of repression against organized labor. Turning to the displaced, the advent of ultra-surveillance has had a rather surprising implication. It has induced the mirror image of displacement in the form of *bloqueos.* In so many ways, surveillance has altered the landscape of spatial politics for social forces.

Overall, PC has relied heavily on military programs. It has largely neglected the important strategic problem of reconstructing the Colombian state in an attempt to win the hearts and minds of the population. While a more orderly environment has been enthusiastically welcomed during President Uribe's reign, the material interests of the majority population have not been satisfied by PC. As such, PC represents a myopic formula for defeating the Left militarily and for promoting foreign economic interests. This argument will be developed further in the following chapter, which will examine other key aspects of the RMA in relation to PC, such as the privatization of warfare and the nexus of crime, war, and terror.

4

Colombia and the Privatization of Warfare

Colombia's legacy of violence, fragmentation, and a failed state was examined in Chapter 2, and serves as a backdrop from which to consider social forces in relation to PC and the RMA. The themes of terror, asymmetry, complexity, and ultra-surveillance, as well as the interaction among these themes for a synergetic effect, were explored in Chapter 3. Here we shall address a crucial component of the RMA that interacts with the others we have treated so far—the privatization of warfare and associated phenomena. Featured will be an analysis of privatized security in the form of private military corporations and private contractors, or what some deem to be neomercenaries. Next, the link between crime, war, and terror will be analyzed in relation to asserted social forces. Also considered will be the social effects of privatizing state corporations. There will be a special emphasis on two significant components of Colombian civil society in relation to PC and the RMA—these are organized labor and the displaced population. This chapter will conclude with a general assessment of PC and a discussion of how the relation between various components of the RMA explains strategic change in the country.

The State, Corporations, and Private Security

Private security is a key component of the RMA, as we observed in Chapter 1. Private *security* corporations have performed a crucial role in the Colombian imbroglio. Because of the nature of conflict in the country, it is sometimes difficult to distinguish between such businesses and private *military* corporations, as private security often relies on military or police techniques. As the line blurs between defense and protection, on the one hand, and offense and combat, on the other, so too does the distinction between private security and private military corporations. Further, due to the controversial nature of such companies, they try as much as possible to operate clandestinely. They are on record as operating in Colombia since at least the 1980s, when an Israeli business, Spearhead Limited, trained members of Colombian drug cartels led by Pablo Escobar and José Gonzalo Rodríguez Gacha.[1] This marked a shift for the cartels, which had previously leased FARC members to defend their business interests. But that alliance proved to be intolerable due to severe ideological tensions.[2]

A significant benchmark occurred in 1992, when British Petroleum (BP) hired Defense Systems Limited, and its Colombia branch, Defense Systems Colombia (DSC), in a 1-million-pound deal to protect its interests in Casanare. At the time, Casanare represented South America's biggest oil find in 20 years. A person identified as a former security adviser to DSC is on record as stating, "[W]ell, it's secret, but they were providing lethal training to the police."[3] This illustrates the hazy distinction noted earlier between private security and private military corporations. The plan was designed to repel attacks from both the FARC and the ELN, which had repeatedly attempted to tax or extort money from petroleum corporations, threatening to bomb infrastructure or to kidnap executives. These threats were real enough, as exemplified when the ELN in October 1998 blew up a BP pipeline in the village of Machuca, Antioquia, killing 70 people.

There have been other notable controversies regarding DSC, including allegations that it provided counterinsurgency training to the 16th Brigade of the Colombian army and also supplied the military with intelligence data regarding those who opposed projects operated by BP.[4] Related to this, it is clear that BP signed an agreement to assist the 16th Brigade by providing security and communications equipment, administration materials, information, engineering and health services, helicopter time, and land transport.[5] Subsequent to that agreement, some protesters to BP's operation were assassinated and many were threatened.[6] Further, a respected academic suggested that DSC was hired along with an Israeli corporation, Silver Shadow, to defend BP's oil interests in Antioquia and Arauca, and that this coincided with the paramilitarization of Arauca.[7] Allegations regarding any link with paramilitaries have been repeatedly denied by DSC officials, who have emphasized that no evidence exists to support such charges despite governmental inquiries into the matter.

More recently, the major American private military corporation MPRI was awarded a $4.3 million contract in 1999 by the U.S. government to provide advice as to how to structure warfare in Colombia. MPRI officials were located inside the base of the Colombian Armed Forces High Command along with U.S. defense personnel.[8] This contract was awarded during the year that PC had been formally announced, prior to its enactment in 2000. It is likely, then, that MPRI was hired by the U.S. government to provide guidance regarding the formulation of PC. Thus, not only has PC relied heavily on private military corporations, but the plan itself was also likely devised based on counsel from a major private military corporation. In 2000, MPRI was awarded a $6 million contract to provide training and advice to the Colombian military as part of PC. MPRI employees involved in that project included former members of the U.S. military and the CIA. The corporation had at least 16 employees working on the Colombian project prior to the enactment of the training program.[9]

Existing private military and security corporations were joined by a barrage of newcomers associated with PC, and this became increasingly apparent by 2002 and 2003. In 2002, Lockheed Martin provided radar systems associated with PC, and Sikorsky Aircraft as well as Bell Helicopter Textron supplied fighter helicopters. By 2003, the U.S. private military corporation Dyncorp was estimated to have at least 44 permanent staff and 65 rotating employees in Colombia who flew U.S. helicopters and planes for their mission.[10] Others involved in PC by 2003 included Arinc, which provided training, equipment, and intelligence associated with the fumigation of coca crops; TRW, which supplied radar systems; Matcom, which provided logistical coordination services for U.S. and Colombian military personnel; Air Park Sales, which offered equipment for riverine battle; Integrated Aero Systems, which provided aircraft and communication devices; and California Microwave Systems, which supplied fumigation pilots.[11] The U.S. Department of State indicated that by 2003 private contractors from at least 16 U.S. companies were present in Colombia.[12] Finally, by the end of 2003, President Uribe urged the estimated 170,000 employees of private security and military corporations to provide intelligence to the Colombian government.[13]

There are some important but unanswered questions regarding the utilization of contractors in Colombia. For example, a particularly murky issue concerns the maximum number of contractors permitted in Colombia at any time. When PC commenced, the official limit of U.S. contractors was set at 400, and this was expanded to 600 in 2004. But these limits only referred to U.S. citizens, and there was no limit on the use of contractors from other countries. Thus, there could have been thousands of private military employees working in Colombia at any point during the PC era. A cable news network reported having observed 1,000 contractors in a single location during its limited visit to Colombia in February 2000. There has been no agency in place to monitor their actual number and it would be difficult in many cases for observers to distinguish contractors from military members. There have also been questions regarding the full range of services offered by private military and security corporations. Part of this has to do with the secrecy that shrouds them. One NGO, for example, reported in 2004 that none of its requests regarding questions associated with the use of U.S. contractors in Colombia, submitted under the U.S. Freedom of Information Act, have been fulfilled.[14]

Some companies, however, have been forthcoming regarding private security services as well as the security predicaments faced by transnational corporations (TNCs) operating in Colombia during the PC era. The manager of security for Nexen, a Canadian oil company operating in Colombia, indicated that his company obtained security through a legal arrangement with the Colombian military. Nexen had essentially leased members of the Colombian military by paying the Colombian government $60 per day, per soldier. Associated problems included the military's occasional abandonment of services to Nexen when unpredicted security crises arose. Further, some locals in areas where Nexen operated indicated that they feared that military protection of the oil site could leave them vulnerable to guerrilla attacks when the oil site was shut down or when the military were not present. Nexen provided electrified gates, surveillance devices, sensors, and armed cars as part of its security operations, and also attempted to limit as much as possible the placement of Canadian executives in the field. Finally, Nexen denied that it made any payment to the guerrillas to offset the possibility of attacks.[15]

The director of information for DSC, the security corporation mentioned earlier that was engulfed in controversies in the 1990s, indicated that during the tenure of PC the company has offered distinct menus of services for clients in large urban centers versus those in the countryside. With regard to urban clients, such as those in the pharmaceutical industry, the company typically has provided protection against kidnapping and theft.[16] DSC has also offered its services to enhance security at airports throughout Colombia, ranging from bomb protection to prevention of hijacking. While it offered advice on how to structure and implement security at airports, it has not provided security personnel.[17]

It is in the countryside, where DSC's clients operate in the extractive sector, that DSC has met controversy. A DSC official indicated that his corporation does not itself use arms, but has contracted out to other private military groups that are legally permitted to do so. In essence, then, he indicated that DSC has worked with private armies. This service has been necessary, he explained, because the Colombian military is not sufficiently strong to protect TNCs against guerrilla attacks. Further, he indicated that his company has provided to its clients in the extractive sector electronic surveillance equipment, including cameras, sensors, infrared night vision, motion detectors, and microphones, and also helicopters, armored cars, and escorts for executives.[18]

There are a number of debates surrounding the use of contractors, private security companies, and private military corporations. One of these concerns the question of accountability, as we saw in the last chapter regarding the Airscan case. Who is responsible when problems arise—individual contractors, the private company for which they work, and/or the U.S. government and military? Who decides who is accountable? Who punishes those found accountable? Surprisingly, there has been no durable debate in either Colombia or in the United States regarding this crucial problem.[19]

Also important is the issue regarding the safety of contractors from the United States and other countries.[20] Three U.S. contractors were kidnapped by the FARC in February 2003, and had not been freed by mid-2007. Further, at least 11 contractors were killed by the middle of 2004, including fumigation pilots who lost their lives in air crashes that were likely caused by rebel attacks. For its part, the FARC in 2004 disdainfully observed the strong growth of private defense industries in Colombia, portraying them as "criminal businesses" designed to protect the interests of the wealthy and of TNCs. The rebels viewed this as another form of "paramilitarization."[21] It is clear that the FARC and the ELN view contractors as their enemies, thus placing them in extreme danger. Perhaps because of mounting deaths and kidnappings, for the first time since 2000 the United States in fiscal year 2005 reduced "by more than 100" the number of contractors engaged in its antidrug operations.[22] What safeguards exist for contractors? Does the U.S. government work as hard to protect the security of contractors as they do for U.S. military members and government officials? These questions remain unanswered.

Despite the serious problems associated with the use of contractors, there are several factors that propel their extensive use. They serve a number of objectives for the U.S. government. They are hired on a limited, contractual basis, and so Washington is not committed to a four-year term as it would be with official military personnel. Further, contractors do not get the benefits enjoyed by the military, in terms of the provision of education, pension plans, and the like, once their tenure is terminated. Also, some consider that contractors are more efficient, as they can be specially selected to perform particular duties. For example, racial background, language ability, and other factors can render certain contractors less obvious, and therefore less vulnerable, in the battlefield. Private military corporations such as MPRI and Dyncorp possess sizeable databases that permit them to deliver tailor-made private military personnel. Also, the use of contractors may render U.S. involvement in places like Colombia less important to the U.S. media and to the American public than if official military members are used. Finally, the privatization of warfare fits into a larger ideological framework that supports privatization across the board—in education, health, welfare, pension funds, and so on.

Crime, Terror, and War: Actors and Interests

> For him there was no such thing as good or bad, clean or dirty money.[23]
>
> *Fernando Vallejo,* Our Lady of the Assassins

Widespread private financing for war is a distinctive feature of the RMA. Although it may appear to be a recent phenomenon in many parts of the world, it is more accurately the reemergence of a feature that was dominant prior to the advent of the modern nation-state. In the case of Colombia, which never achieved a modern state, privatized war is nothing new. From the interminable feuds between Liberals and Conservatives from the 1830s through the 1950s, to the armed clashes between rival mobs over assorted contraband, privately funded warfare has been a persistent feature in the country. Its presence became exaggerated in a trajectory beginning in the 1980s, and has been fueled principally by narcotrafficking, and to a lesser extent by extortion, kidnapping, counterfeiting, and other types of crime. The purpose at hand will be to draw the links between crime, Colombian warfare, and the politics of fear. While the general topic of narcotrafficking has rightfully received considerable academic attention,[24] the particular focus here will be on its relation to the privatization of war. Let us begin by examining the motives of relevant actors involved in the Colombian imbroglio in relation to narcotrafficking.

For the United States, the central problem is that the illicit drug trade has greatly empowered what has become the most militarily powerful leftist insurgent group in Latin America, the FARC. Colombia is the source of over 80 percent of the cocaine and about half of the heroin on U.S. streets. *Janes Intelligence Review* estimated that the FARC received about $300 million in 2004 from narcotrafficking, while the Conservative Colombian newspaper *El Tiempo* suggested that the FARC's 2003 income from all sources was approximately $77.16 million.[25] While the exact profit the FARC reaps from its involvement in the drug trade is uncertain, what remains clear is that narcotrafficking has subsidized the group's substantial military machine. The central interest of the United States with regard to Colombia's illicit drug trade, then, has been to diminish the FARC's access to it, thereby reducing the military and financial power of the rebels.

To achieve this objective, the United States has relied on three principal strategies: massive aerial fumigation of coca crops, the shooting down of suspected air shipments of cocaine in tandem with other interdiction efforts, and extradition of suspected drug kingpins to the United States. During the period between the inception of PC in 2000 and the end of the year 2006, the United States sprayed the herbicide glysophate, which is designed to kill all plant life with which it comes in contact, on at least 1.37 million acres of land suspected of coca cultivation and an additional 52,000 acres suspected of opium growth in Colombia.[26] Along with Plan Patriota and the restructuring of the Colombian military to better fight the FARC, fumigation has been central to U.S. efforts with respect to PC. It has been complemented by the manual destruction of coca plantations in Colombia's national parks, whereby 28,240 acres of coca bushes were eradicated by November 2005.[27] Further, since August 2003, the United States has supported the Colombian practice of shooting down suspected drug flights. Finally, an important U.S. weapon with regard to Colombian narcotrafficking has been the extradition of suspected drug kingpins. The first FARC guerrilla was extradited to the States on drug charges in 2002, and there has been a steady parade since then. Among the most salient cases is the extradition on 31 December 2004 of Simon Trinidad, or Juvenal Oviedo Ricardo Palmera Pineda, a senior FARC member who was among the group's top financial strategists. The U.S. Department of State deemed him to be a "narco-terrorist."[28] In March 2006, Washington charged 50 leading members of the FARC with narcotrafficking, and formally requested their extradition.[29]

The interests of the Uribe government in Colombia largely echoed those of the United States. Although the Uribe administration embraced massive amounts of U.S. military assistance, it faced obvious contradictions vis-à-vis narcotraficking. An AUC leader, Vicente Castaño, stated in 2005 that the paramilitaries controlled 35 percent of the Congresspersons in Colombia—a figure that by late 2006 seemed to be a tad low in the context of a scandalous demobilization process that exposed alarming ties between the government and the *paras*—a point to which we shall return. AUC leaders have admitted over the years that at least 70 percent of their funding is derived from illicit drug trafficking. There have been persistent allegations against the unhealthy closeness between paramilitaries and Uribe, during both his tenure as president and as governor of Antioquia.[30] Moreover, a declassified U.S. intelligence report dated 23 September 1991 indicated that then-senator Uribe was "dedicated to collaboration with the Medellin cartel at high government levels" and that he was a "close personal friend" of Pablo Escobar—an allegation that was energetically dismissed in 2004 by both the Colombian and U.S. governments.[31] Given Uribe's less than hostile arrangement with the drug-smuggling paramilitaries, it is highly possible that his government has appreciated the fallacy of U.S. efforts to reduce narcotrafficking in Colombia but has nevertheless placated Washington in its efforts to obtain military assistance to combat the FARC and the ELN.

Finally, the FARC and the AUC have witnessed both common and competitive strategic interests vis-à-vis the illicit drug trade. In one sense, their motives have been identical: to promote narcotrafficking in order to finance their military power and political capacity. Despite their animosity on so many fronts, there have been claims that the FARC and the AUC have cooperated in a limited way when it comes to sustaining their mutual interests in the illicit drug trade—particularly regarding drug flows through certain territories.[32] Presumably, though, there has been more conflict than cooperation between them on this front. They have competed vigorously for territory and market share. One example among many occurred in June 2004, when the FARC massacred 34 peasant coca growers, formerly aligned with the AUC, in order to assert FARC control over new territory.[33] There are innumerable other examples to demonstrate that the drug trade is clearly associated with social terror.

Crime, Terror, and War: Strategic Implications

The U.S. war against drugs in Colombia has failed in many respects. It has promoted corruption, ecocide, strategic backfires, and diplomatic embarrassment. Further, as we shall see below, it has not succeeded in reducing the quantity or increasing the price of cocaine on U.S. streets since PC was implemented. Regarding corruption, the United States has suffered a series of embarrassing stings, ones that demonstrate beyond doubt that it is as vulnerable to corruption as its Latin American counterparts. Among the most recent and notable cases was the conviction of the U.S. drug czar James Hiett in Colombia, in a 1999 case that featured his role as money launderer for the proceeds of his wife's trafficking of cocaine through U.S. diplomatic mail. Colonel Hiett got a five-month sentence, his wife got five years. U.S. contractors also have been implicated in cocaine trafficking, as have U.S. soldiers. For example, five U.S. army soldiers were arrested in March 2005 for smuggling 35 pounds of cocaine from a Colombian army base into El Paso, Texas.[34] U.S. army officers were arrested in May 2005 for attempting to sell 40,000 rounds of ammunition to the AUC, which the AUC presumably was able to purchase through its role in narcotrafficking. These cases triggered an outcry of hypocrisy and double standards in Colombia after the soldiers received immunity from Colombian prosecution upon insistence by the United States to respect the guidelines set forth in a 1962 agreement between the two countries.

This brings us to an assessment of U.S. extradition. It is a feared instrument of U.S. policy among Colombian narcotraffickers and insurgents. Approximately 450 persons were extradited to the States between August 2002 and December 2006.[35] The U.S. justice system is viewed as being far harsher than Colombia's, where justice officials are routinely assassinated, intimidated, or bought off and where the prison system can be manipulated by powerful inmates. It can represent a strong bargaining chip for Washington, especially in the context of negotiations with insurgents. There are also important limits to this policy. Capturing leading insurgents and narcotraffickers is far harder than extraditing them. Further, extraditing drug traffickers will not reduce the flow of the trade, as others will rush in to fill the void. Thus, this policy cannot address the root of the problem. There is also an obvious double standard with regard to the targets of extradition. So far, the United States has been keenly interested in extraditing leading FARC figures and disinclined to extradite members of its strategic ally, the paramilitaries.

The biggest and most pernicious failure of PC has been the U.S. reliance on biological warfare through fumigation. The number of hectares sprayed with glysophate increased steadily every year—47,371 hectares in 2000, 84,231 in 2001, 122,695 in 2002, 132,817 in 2003, 136,555 in 2004, 138,795 in 2005, and 160,000 in 2006. Yet, according to figures released by the U.S. Office of National Drug Control Policy, cocaine was cheaper and the quality was purer in 2003 than when PC began in 2000.[36] Further, cocaine prices on the U.S. streets averaged about $200 a gram with about 60 percent purity in 2003, falling in price to $140 a gram in 2006 and rising to 70 percent purity in that year. Corroborating that, figures released by the UK government show a steady drop in street prices of cocaine from 2000 through 2005.[37] Thus, the supply of cocaine appears to have grown despite Washington's drug war. Moreover, satellites from the United Nation's Office on Drugs and Crime noted that 44 percent of coca crop locations in 2005 did not exist in 2001–2004, and that the overall crop was burgeoning. The U.S. plan did not work for a simple reason: the strong incentive for sustained production given the huge profits derived from the illegality of the trade. Coca growers and producers compensated for fumigation through the balloon effect, whereby cultivation spreads to other areas, and by the use of fumigation-resistant plants with increased yields, smaller and dispersed crops, crops hidden under banana trees, and so on.

Beyond failing to achieve its objective, fumigation has produced a series of highly negative consequences. Strategically, it pushed *cocaleros*—peasants who grow coca as a subsistence crop in the context of a broader agricultural crisis—closer to the FARC and AUC, who defend their meager livelihood. Alternative development programs paired with fumigation generally have failed, fueling the perpetuation of the illicit trade. For example, the General Accounting Office of the U.S. Congress observed that alternative development programs "benefit very few persons," suffer from "problems of implementation," and are "not sustainable."[38] There is also the question of fumigation being toxic to humans and animals.[39] For example, it is generally required that people not enter sprayed areas for 12 hours after spraying, but there is questionable enforcement of this. Further, fumigation planes, to control the target area, are meant to spray 10 feet above crops, but often must fly higher to avoid armed attacks, thereby causing the spray to drift.[40] As the herbicide is designed to kill all plant life with which it comes in contact on the ground, there are ripple effects among animals and other elements of the local ecology. The process has placed in danger the lives of contractors who perform such duties, as was noted in the previous section. Fumigation along Colombia's border with Ecuador ignited a feud in late 2006 and early 2007 with the new leftist president of that country.

Given the disastrous and expensive legacy of the drug war in the Andes and elsewhere since at least 1989, all of the failures regarding PC's fumigation process are predictable and represent lamentable re-runs of past mishaps. Perhaps most striking about all of this is that the lion's share of funds allocated to PC has been devoted to Narcotics Control Assistance (NCA) (see Table 3.2 in Chapter 3). NCA made up 99 percent of U.S. military aid to Colombia in 2000, 87 percent in 2001, 91 percent in 2002, 94 percent in 2003, 79 percent in 2004, and about 81 percent in both 2005 and 2006. To the extent that it was meant to reduce narcotrafficking, the policy has clearly failed. On balance, however, we noted in Chapter 3 that NCA has meant far more than just fumigation. It also provided for the restructuring of the Colombian military to better fight the FARC within the context of asymmetric warfare. To that extent, NCA may have met with limited success—a point to which we shall return. But, overall, many key components of the U.S. drug war have not been successful.

The United States is not the only actor tarnished by the drug war. The Colombian government, including the police and military, has suffered countless scandals involving the corruption and debilitation of an already anemic and inept state. The narco-cassette scandal of the 1990s, which suggested hefty drug payoffs to President Samper, and a later episode involving the "disappearance" of four tons of cocaine seized by government forces, are among the most sensational cases of what is clearly an immensely corruptive influence emanating from narcotrafficking. And while the FARC and AUC have reaped huge profits from the drug trade to bolster their military power, they have rendered themselves vulnerable to charges of illegitimacy. Can the FARC make a convincing claim to harbor Marxist and egalitarian ideals when it has enriched itself through illicit transnational capitalism to fund its brutal military machine? Can the AUC establish legitimacy, domestically and globally, when leading members of this ferocious group admit that drug money has helped them purchase a third of the Colombian Congress?

This same general theme is true with regard to the defamatory effects of other types of crime. The FARC and the ELN, for example, are estimated to have kidnapped 15,392 persons in Colombia between 1996 and 2002,[41] and the government estimated that about 2,000 kidnapped persons remained in captivity by the beginning of 2004. And while Mexico edged out Colombia with regard to the number of kidnap victims in 2006, Colombia remained in the lead on a per capita basis. Further, Colombia has remained the world center of counterfeiting, with about 50 percent of total global counterfeiting estimated to occur in Cali. In 2004, the FARC was caught in a sting when it attempted to buy $2.9 million of weapons with false currency. To gain popular legitimacy, and to be a credible political leader rather than just a criminally based military machine, insurgent groups must present themselves as moral actors pursuing popular ideals. They must become social heroes of sorts. But narcotrafficking and other crimes in Colombia have brandished those pretending to be good guys.

Consumption, Colombia, and the RMA—from Drugs to Oil

Among the reasons for the resounding failure of the U.S. drug war has been its emphasis on the forces of production rather than of consumption. Clearly, a focus on production is crucial when examining the politics of class forces and patterns of exploitation, and in terms of analyzing what is produced within the context of a particular mode of development. But in the Colombian case, a focus on consumption is also key, especially with respect to Northern consumption and vertical dimensions of power. The U.S. drug war in Colombia failed principally because it attempted to eradicate production without appreciating the driving force of Northern consumption.[42] This omission is strange, especially among neoliberals who purport to understand the inner workings of capitalism, because it ignores the classic economic equation of supply and demand by concentrating only on supply.

More broadly, it is worth underscoring that the latest RMA has been driven to a considerable extent by patterns of U.S. and Northern consumption, especially regarding oil in relation to Middle Eastern politics. Similarly, the manifestation of the RMA with regard to PC has been driven first by U.S. consumption of cocaine, and since 9/11 and especially the invasion of Iraq, by American oil interests. Indeed, in 2001, the Bush administration's National Energy Policy observed that Colombia "is becoming an important supplier of oil to the United States."[43] Colombia has been important to U.S. oil interests for two broad sets of reasons: the oil Colombia possesses, and the country's role as a bastion of U.S. political and military support in an oil-rich region that is generally hostile to American interests. During the beginning of the new millennium, Colombia has been the tenth largest supplier of oil to the United States. Petroleum represented about 27 percent of Colombia's legal exports in 2005. Oil revenues rose 80 percent in 2006 over 2004 figures. While Colombia's annual production of oil has decreased in recent years, from 844 million barrels in 1999 to 527 million in 2006, 80 percent of the country's territory has not been explored. The country's state oil corporation, Ecopetrol, suggests that Colombia possesses huge reserves ripe for exploration, with a potential of 47 billion barrels. Exploration has been proceeding apace, exemplified by Oxy's $30 million project during 2004–2008. Colombia's proven reserves of 1.8 billion barrels represent about a third of Venezuela's, the fourth largest exporter of oil to the United States.

Given the increasing preciousness of petroleum, in the context of Middle Eastern conflict and growing consumption in key countries such as China and India, Washington's strategic interest has been to protect existing supplies of Colombian oil and to promote exploration to increase future reserves. The threats to Colombian oil are clear enough. Oxy's Caño-Limón pipeline was bombed over 1,000 times between 1990 and 2003, 178 of which occurred in 2001, leading to $500 million in losses that year alone. Explosions in the Oxy pipeline have declined to a relative trickle since the implementation in 2002 of U.S.-led security measures. In that year, PC allotted $99 million to protect the Caño-Limón petroleum pipeline, one of five in Colombia, in an operation involving scores of Green Berets tasked with training Colombians to protect the pipeline in the future. Similar annual allotments have continued since, such as the $90 million granted for such purposes in Fiscal Year 2005.[44] There were only 20 bombings of the pipeline in 2005, and sporadic bombings in 2006 including a major attack by the FARC just before the May 2006 presidential elections. A nearly 50 percent reduction in the special U.S. forces, which comprised the Green Berets, occurred in October 2005, when 150 of its members were diverted from Colombia to Iraq. By 2007, it appeared that Colombian forces were sufficiently trained to fill the gap left by departing U.S. troops.

Limited military expenditures also have been divulged to other areas where petroleum is being produced, such as Putumayo, which has been highly vulnerable to attacks. The FARC has launched armed strikes against petroleum infrastructure there twice in 2002, 23 times in 2003, and 24 times in 2004. During its "comeback" in June 2005, the FARC attacked the main military base guarding petroleum installations in Teteyé, Putumayo, as well as the petroleum plant itself, killing 19 soldiers. Thus, not only have the guerrillas identified oil installations as strategic targets, but also this particular episode may have involved up to 500 FARC troops, exposing seriously weak links in the government's military surveillance and intelligence systems.[45]

* * *

The United States, through its own initiatives and through its influence in the IMF, has encouraged sweeping privatization in Latin America since the debt crisis of 1982. This has served the interests of U.S.-based and other TNCs. The policy has been manifested in Colombia with regard to a wide variety of formerly state corporations, in areas such as telecommunications, utilities, education, health, and so on. The dynamic here is noteworthy. In 2005, for example, to ensure eligibility for future loans, Colombia was instructed by the IMF to lower its public debt from 55 to 50 percent of its GDP by the end of 2006.[46] So, in the context of a civil war, rising military spending, and a state that is absent in much of the country's territory, the IMF instructed Colombia to cut social programs. Clearly, this was strategically unwise. But the IMF offered what it viewed as a solution: to make up for cuts in social spending, sell off state corporations to TNCs and use that money for social programs. While this plan benefits TNCs, it may not be in Colombia's long-range interests.[47]

In Colombia's case, Ecopetrol develops 14 percent of the country's petroleum, with the rest developed by independent companies or partners of Ecopetrol. Colombia's key petroleum labor union, USO, conducted a major strike in spring 2004 to protest what it called the imminent privatization of Ecopetrol. Despite its denial at the time of any privatization scheme, the government decided officially in 2006 to privatize 20 percent of Ecopetrol. This general situation has numerous implications for Colombia's security situation, in terms of labor unrest, assassination and intimidation of union members, strategic overlap between unions and leftist guerrilla forces, and so on. Here we confront not only the privatization of warfare but also how privatization at the behest of TNCs can fuel warfare and be a central point of contention upon which strategic conflict is based.

There is a clear regional dimension to all of this. More than two decades after the policies of neoliberalism were implemented, a serious backlash has occurred in Latin America, as evidenced in Argentina, Brazil, and Uruguay, and with regard to the politics of oil, in Venezuela, Ecuador, and Bolivia. In the context of popular comparisons between the current race to exploit oil and the raiding of gold and silver during the conquest, Latin American petroleum and gas increasingly have been viewed by locals as precious strategic resources that can promote national development and social welfare.

Related to this have been strategic tensions between the United States on the one hand and Venezuela and Bolivia on the other. Colombia has not only sided with the United States, but has also been engulfed in its own disputes with the Chávez government. For instance, in a bizarre display of its penchant for the privatization of warfare, Colombia's Uribe administration hired bounty hunters in December 2004 to enter Venezuela illegally in an effort to track down FARC leader Rodrigo Granda and bring him back to Colombia. Colombian vice president Francisco Santos stated, "Hopefully all the bounty hunters of the world will come here to capture these bandits. The money is here for them and the rewards are good and can be handed out anywhere in the world."[48] Beyond the tensions with Venezuela, the nationalization of the gas and petroleum industry in Bolivia has been perceived as a challenge to the interests of the United States. Against such a backdrop, Colombian oil is more important than ever to Washington. Further, Colombia may serve as a site from which to project U.S. interests in Venezuela, Ecuador, and Bolivia. In many ways, U.S. strategic interests have shifted from drugs to oil during the tenure of PC.

Social Forces, Privatization, and the RMA: Organized Labor

> They think everything's the guerrillas, and you never know what they'll do.[49]
>
> *Álvaro Mutis,* The Adventures and Misadventures of Maqroll

Colombia's labor unions are a crucial component of the country's social fabric, because, as noted, they historically have been among the best organized elements of civil society. And they clearly have been under siege, with privatization representing their biggest threat. Colombian unions have found themselves challenged by both national and global forces. Nationally, they have been viewed with hostility by a trinity comprising the government, large business owners, and paramilitary groups, who together portray organized labor as an ideological enemy with links to leftist guerrillas. Globally, Colombian unions have faced threats from TNCs, many of which prefer nonunionized labor in efforts to reduce wage and benefit costs. Labor also has faced pressure from the IMF's neoliberal policies that encourage the privatization of state corporations in the context of a debt crisis—Colombian labor unions have called for a moratorium on the country's foreign debt.[50] Further, Colombian unions have not been greeted warmly by the U.S. government, which has resented them due to their antipathy to hemispheric free trade agreements.

In a country where civil society clearly needs to be substantially fortified, it is deeply worrisome that one of its essential components, labor unions, has been depleting rapidly. In the 1980s, about 15 percent of Colombia's active working population were union members. By 2004, the figure fell to just 4 percent.[51] Moreover, the number of newly registered unions dropped dramatically from 195 in 2001, the year prior to Uribe's election, to a meager 14 in 2005.[52] About 50,000 union jobs disappeared between 1990 and 2004.[53] This has been the result of a number of factors. These include an economic crisis in the late 1990s and in the early part of the new millennium that shed well-paying union jobs, a sharp tendency toward more "contract" or nonpermanent work, as well as more "informal work," and so on.[54] Within such a context, privatization has represented the biggest threat to unions, because it typically means the loss of union jobs, either through the disappearance of positions or through the hiring of nonpermanent and nonunionized labor.[55] For example, when Colombia's Telecom was privatized in 2003, half the

Table 4.1 Human rights violations against Colombian union members, 2000–2005

	2000	2001	2002	2003	2004	2005
Threats	180	234	190	296	445	260
Assassinations	135	198	184	91	94	70
Disappearances	17	12	9	6	7	3
Detentions	37	8	11	49	77	56
Hostage taking	2	13	19	54	17	32
Kidnapping	24	41	27	7	5	6
Displacement	155	70	2	91	33	8
Torture	2	1	1	0	1	1

Source: Statistics from Escuela Nacional Sindical, "Informe Sobre la Violación a Los Derechos Humanos de los y las Sindicalistas Colombianos," Medellín, April 2006.

employees lost their jobs, and many of those hired back were offered temporary, nonunionized contracts.

Table 4.1 demonstrates the truly horrific levels of human rights abuses suffered by Colombians who have been pressing for basic labor rights. While the decrease in homicides under the Uribe government is a hugely important step in the right direction, murder rates of union members have remained astronomical by global standards. Although there have been fewer assassinations, other abuses have risen. More threats, selective assassinations, detentions, and hostage taking suggest that organized labor has remained under siege, and that it has been the target of a refashioned array of tactics. For example, during the summer of 2006, the leader of USO Cartagena received death threats presumably from paramilitaries against him and his family, and was forced to leave the country.[56] This kind of high-profile targeting has a compensatory effect for fewer union members murdered. Assassination of union members rose in 2006 from the previous year to 84.[57] Still, union leaders are indeed grateful for a generally declining trend of assassinations against their members since 2002, an accomplishment they attribute to their efforts to attract greater international attention to the labor situation in Colombia.[58] Overall, repression aimed at union members has been a deeply worrisome feature during the tenure of the Uribe administration and of PC. It signals an attempt to erode through force and fear the meager progress achieved for basic workers' rights in a country considered the most dangerous in the world to be a union member.

Let us examine the plight of particular unions in Colombia and their relation to warfare in the country. A classic and high-profile case involves workers at Coca Cola bottling plants in Colombia. In 1992, 96 percent of Coca Cola workers were full time and unionized, which dropped to 4 percent by 2004.[59] Some of the workers went on a hunger strike in March 2004 to protest the murder of nine of their union members and death threats against 67, in addition to a host of other abuses.[60] Given Coca Cola's high-profile status as the epitome of U.S. business, it has contributed to a perception among some members of Colombian labor that U.S. TNCs are antiunion and may engage clandestinely in repressive activity. Further, the president of Colombia's second largest umbrella union, the 350,000-member CGTD, suggested that the trend to shrink the union movement commenced in Colombia during the 1990s. That decade introduced the neoliberal policies associated with the Washington Consensus, he said, adding that antiunion policies have been operationalized through the World Bank and the IMF to serve transnational corporate interests.[61]

Also important is the example of the CUT, the country's biggest labor umbrella organization, with 750,000 members. A CUT spokesperson suggested that unions have become targets under the Uribe administration, because they oppose the president's agenda of privatization, particularly with regard to the sectors of health, education, and telecommunications. Further, some key union leaders and members are suspected by the government of supporting the FARC and the ELN. Hence, unions are viewed as the strategic enemy in the context of internal warfare and the implementation of PC. CUT leaders assert that PC has meant less opportunity for free expression by Colombian unions. This is due to increasing U.S. corporate interests in the country, which clash with union agendas, related threats and detentions vis-à-vis union rank and file, and the government's penchant for falsely equating antiprivatization and anti-neoliberal policies with guerrilla membership or support.[62] Thus, key components of a strong civil society—free expression and nonviolent conflict resolution mechanisms—are debilitated in the context of warfare and the imposition of a transnational corporate agenda.

The plight of Colombia's major oil union, USO, demonstrates a similar tale of struggle in the face of repression and looming decimation. USO had 13,500 members in 1994, which depleted to 2,350 by 2004, with permanent unionized employees discarded for nonunionized contract workers. Since 1986, 108 USO members have been assassinated and another 300 have had to leave their jobs due to death threats. USO leaders say paramilitaries are responsible for most of the attacks and for the campaign of terror launched against them.[63] USO leaders point to a clear transformation associated with PC, particularly after the September 11 attacks. The post-9/11 environment coincided with an increasing empowerment of paramilitary forces in key Colombian oil-producing regions, such as Barrancabermeja and the department of Arauca.[64] The paramilitaries have mixed terror with economic enticement to exercise social control in these regions. In relation to this, an astonishing 5 percent of Colombian oil is allegedly "stolen." USO members and others contend that TNCs turn a blind eye that allows paramilitaries to "steal" the oil and resell it to the local population at cut rates in order to generate community support for the AUC.[65]

Despite such tactics of the paramilitaries, USO has been able to generate crucial social support in key oil-producing cities such as Barrancabermeja, where it prompted 35,000 residents to take to the streets in support of USO's pivotal 2004 strike against what it claimed was the impending privatization of Ecopetrol. The strike, which lasted almost two months, was declared illegal by the Uribe government, which deemed oil production to be an "essential service." As a result, 263 USO members were initially fired, including key strike leaders and seven of its national board of directors. Throwing its full support behind USO and its legal right to strike, the International Labor Organization (ILO) issued a statement in May 2004 that was highly critical of the Uribe administration and that emphasized that the Colombian government had the worst record of labor rights protection in the world.[66] The government was told to rehire the employees it illegally fired, but by May 2005 34 of them still had not been hired back despite a signed agreement between USO and Ecopetrol.

Like their counterparts in Venezuela, Ecuador, and Bolivia, USO members have demanded the nationalization of the Colombian petroleum industry. But USO leaders recognize that Colombia, in contrast to these countries, has witnessed a climate of intense intimidation, terror, exclusion, and fragmentation that has profoundly limited the influence of unions. From their perspective, PC has served to strengthen the paramilitaries and transnational corporate interests. It has meant the detention and harassment of union supporters under the false guise of links to guerrilla movements. Unsubstantiated media reports suggested, for example, that USO leader Gabriel Alvis had ties to the FARC—an allegation he angrily denied.[67]

Another noteworthy aspect of PC, one broadly related to the wave of privatization due to its link with neoliberal economic policies, is its parallel relationship to the establishment of a free trade agreement between Colombia and the United States. Colombian unions have been critical of such an agreement, as they have been with regard to a hemispheric free trade zone, because they fear additional losses of unionized jobs. During President Bush's trip to Colombia in November 2004, wherein President Uribe called up no less than 15,000 troops to protect him, Bush linked PC to prospects of free trade with Colombia.[68] This has not gone unnoticed by Colombian belligerents. For example, the ELN sees a connection between PC and the Inter-American free trade agreement ALCA (Area de Libre Comercio de las Americas), while the FARC has issued statements suggesting that PC is the military arm of ALCA.[69] A free trade pact between the United States and Colombia was being debated in the U.S. Congress in mid-2007.

Overall, there is a clear link between PC and the struggle of organized labor. It is highly positive that fewer union leaders have been assassinated during the plan's tenure. This is likely due to a growing sensitivity on the part of both the U.S. and Colombian governments regarding horrendous human rights abuses in the face of rising international criticism. But other abuses increased during that era, especially detentions, and also selective assassinations, which involve high-profile targets. These limit the number of cadavers, but transmit through terror the same political message. Overall, tactics have changed, but the message is the same: you may have to pay with your life if you wish to struggle for workers' rights. This message was underscored when Colombia's vice ministro de Trabajo (Labor) told U.S. House of Representatives member Jim McGovern in 2007 that labor unions "inflate" attacks against them, and that "they prosper in blood and violence."[70]

Besides the intense victimization of labor in Colombia, neoliberal restructuring has been associated with a profound contraction of the labor movement since the 1990s, a trend that shrinks one of the country's most organized components of civil society. This spells an ever weaker political voice for those attempting to defend workers' rights from brutal attacks on the part of paramilitaries who serve the interests of private and often international corporations. PC has been implemented alongside newfound U.S. interests in Colombian oil, a U.S. agenda for regional free trade, and growing animosity between Washington and the Andean oil-producing states of Venezuela, Ecuador, and Bolivia. The struggle of organized labor in Colombia demonstrates that it is a country where the distinction between war and politics is blurred to the extreme.

Social Forces, Plan Colombia, and the RMA: The Displaced

> The violence grew worse in the countryside and people fled to the cities, but censorship obliged the press to write about this in an oblique manner.[71]
>
> *Gabriel García Márquez,* Living to Tell the Tale

Nobody has paid a higher price for escalating warfare in Colombia than the country's 3.7 million displaced persons,[72] the largest such population in the hemisphere and the third largest in the world. They have faced loss of homes, property, and livelihood, loss of community and social identity, as well as family breakup—all of this compounded by the psychological trauma generated by the particular cause of displacement. Millions of hectares of land have been abandoned in their wake and absorbed by armed groups in their quest for enrichment.[73] Most displaced never expect to return home.[74] With respect to the displaced population residing in Bogotá, the causes behind violent displacement included threats of violence (57.5 percent), threats of recruitment to belligerents (10 percent), murder of relatives or friends (10 percent), and fear (3 percent), among other reasons.[75] Among displaced children, who make up over half of Colombia's displaced population, 60 percent have witnessed a murder, 78 percent have seen mutilated cadavers, 18 percent have been kidnapped, and 18 percent have seen torture.[76]

The displaced population has faced the ultimate form of political exclusion. They have been perhaps the most terrorized of Colombia's war-torn population. Beyond despair and exclusion, it is crucial to emphasize that they have become dislocated due to fear and terror, so much so that only about 30 percent are estimated to register for assistance with any government agency or NGO due to fear of retaliation.[77] And among those who do accept help, most shun government agencies; over 95 percent of Bogotá's displaced population that have sought assistance have gone to an NGO for help rather than to the government's Red de Solidaridad Social.[78] This is especially true regarding those displaced by the AUC, because they fear that the paramilitaries may sometimes work in concert with government agencies that provide them with information.[79] Members of the displaced population may suffer further exclusion when they relocate, as they may be stigmatized with political affiliations to one of the armed groups depending on their geographical origin.[80]

As Table 4.2 demonstrates, with the implementation of PC, displacement figures rose steadily between 2000 and 2002. Given the commencement in 2003 of the AUC's negotiation with the government concerning their eventual demobilization, combined with the FARC's strategic retreat during much of that year, displacement figures dropped temporarily before rising substantially in 2004 and 2005. Overall, the waxing and waning of displacement figures can be attributed to the intensity of warfare in Colombia. Experts on Colombian displacement suggest, as the director of the NGO Codhes emphasizes, that it is difficult to blame

Table 4.2 Displaced persons in Colombia, selected years

1999	288,000
2000	317,376
2001	341,925
2002	421,553
2003	207,607
2004	287,581
2005	310,387
2006	219,886
1985–2005	3,940,314

Sources: Codhes, "Comportamiento del Desplazamiento 1985–2005," April 2006; Codhes, "2007: Año de Los Derechos Humanos de las Personas Desplazadas," 1 February 2007.

any particular actor for Colombia's displacement predicament, because it depends on the territorial zone and time frame one is considering.[81]

Warfare surrounding oil and drug production is one factor that has contributed to displacement. Massive fumigation of coca crops, or biowar, by the United States under PC is a case in point, as demonstrated in Putumayo during 2000.[82] But the government does not recognize those displaced by fumigation because of the illicit nature of their subsistence farming. While major waves of displacement due to fumigation were predicted in 2000, when PC was first implemented, this did not happen in a prolonged sense because *cocaleros* typically have relocated and continued to engage in coca cultivation.[83] Besides Northern consumption of illicit drugs, the politics of oil consumption since 9/11 is related to the escalation of combat between factions of the FARC and the ELN, on the one hand, and between them and the AUC, the United States, and the Colombian military, on the other. This intensification of warfare surrounding oil production has meant significant displacement in regions such as Arauca since 2001,[84] and in other regions such as Putumayo during 2005–2006.

Colombia's ethnic population has been disproportionately victimized within the context of PC and the RMA. The indigenous population of Colombia, for example, represents less than 3 percent of the total population, yet the Red Cross indicates that over 5 percent of the agency's clients have been indigenous.[85] They have faced increasing displacement from Arauca and other oil-producing departments near the Venezuelan border, where fighting between armed groups has pushed them off their land, and where the development of oil megaprojects has forced their departure. A Guahibo Indian leader in Arauca stated, "When there was no petroleum, there was no war. . . . Before that, we had a dignified life here. Our council of cabildos does not permit them to take the blood (oil) from the earth in our territories."[86] In June 2001, the Latin American Association for Human Rights estimated that half of Colombia's native peoples faced annihilation from encroaching violence associated with land invasion, oil operations, and mega-development projects.[87]

Beyond the sometimes pernicious effects of oil politics, indigenous populations have been routinely displaced from territory considered to be strategically important to armed groups. A classic example is the Sierra Nevada region near

Santa Marta, which is important for coca cultivation and is seen as a strategic corridor for smuggling drugs, arms, and people. Indigenous people in the area include koqui, kankuamo, arhuacos, and wiwa. Further, the native population in this region has suffered from high levels of kidnapping, assassination, and torture.[88] When displaced, the indigenous also lose rights of legal autonomy and self-government over the 28 percent of Colombia's land granted to them in the 1991 constitution. When trapped in *bloqueos,* which means they are forcibly confined to limited spaces, they are denied access to the country's weak array of social services.[89]

Colombia's black population also has suffered disproportionately from displacement, not only in terms of sheer numbers, but in other ways as well. As the victims of the worst racism in the country, they have faced even greater social difficulties than the rest of the displaced population when attempting to relocate.[90] They have been displaced for all the general reasons noted earlier—their location on strategic territory, escalating intensity of warfare, and so on—and increasingly have been displaced due to palm oil megaprojects in the Pacific region of the country. Huge and often illegal palm oil groves have pushed blacks from their land under the enforcement of armed groups, as has been the case in Chocó in 2005.[91] The director of the Asociación de AfroColombianos Desplazados (AFRODES), which represents 800,000 blacks displaced from coastal regions of the Atlantic and Pacific, indicates that displacement, which has worsened for blacks under PC, contributes to the destruction of black culture. In contrast to a sense of community, displaced blacks are dispersed and individualized. As with all displaced families, family breakup becomes more likely, and gender roles change, especially because men typically cannot find work while women can more easily find employment, for example, as domestic servants.[92]

Overall, the militarization associated with PC has contributed to periods of elevated levels of both displacement and confinement of innocent populations. This has been a result of features associated with the RMA such as privatized war, biowar, ultra-surveillance, and so on. While features of the RMA and the intensification of war due to PC are crucial factors that affect levels of displacement, this horrific situation is also the product of another disturbing element. Armed groups in Colombia, on both the left and the right, obviously do not have the sympathy of huge numbers of residents in particular geographic spaces—if they did have their support, residents would likely fight alongside their favored rebel groups as they did in La Violencia and umpteen other feuds that dot Colombia's history, instead of leaving everything behind and fleeing for their lives.[93] Rather than attempting to gain power by relying primarily on political consent from target populations—through the provision of material gains and through the power of ideas—armed groups in Colombia have conceived of power in terms of wealth and military strength so that they can compete with one another as well as with external entities wishing to quell them. Since the 1990s, armed groups in Colombia have viewed territory and resources as their cherished prize, with local populations often viewed as a nuisance that impedes their objective of power maximization. A sad example of this phenomenon was the assassination in January 2007 of Yolanda Izquierdo, the head of an organization in Cordoba that attempted to regain some of the 4.8 million hectares of rich farm land seized by paramilitaries.

Assessing Plan Colombia and the RMA: The FARC

There is no question that PC has weakened the FARC and has placed the rebels on the defensive. Challenged particularly in the countryside, the FARC initially reacted by lashing out through urban terrorism, as we observed in Chapter 3. The culmination of that spree was the horrendous attack on the El Nogal Social Club on 7 February 2003, which killed 36 and seriously wounded more than 100 innocent people who frequented the elite club. Under siege from a barrage of unusually harsh condemnation both nationally and globally for the FARC's apparent role in that bombing, the rebels waited over a month before claiming they were not involved.[94] It is hard to imagine who besides the FARC would have the interest and capacity to bomb the recreational facility in Bogotá's swankest neighborhood, with the apparent message that the war could be brought to the city from the countryside, where it was previously concentrated, and that even super-secure enclaves for the rich are vulnerable to attack. The wave of major terrorist bombings in Bogotá ended with the El Nogal bombing.

Further, alleged links between the FARC and the provisional IRA in the rebel's previous *zona* culminated in August 2001 with the arrest of three IRA members—fortifying speculation that the FARC had received training in urban warfare from members of the notorious Irish rebel group.[95] The FARC has been linked definitively to other acts of terror during this period, but these tended to be outside of the country's major cities. These include the use of cylinder bombs on innocent civilians in Caldona, Cauca, in July 2005 and in Caquetá in March 2006, in addition to a long list of other similar incidents.[96] This general strategy cost the FARC much of the limited international support it had managed to achieve with global actors such as the European parliament and assorted NGOs. It also made the group appear more desperate and brutal than ever to local Colombians.

With urban terrorism proving to be a strategic loss for the FARC, it adopted a policy of relative retreat during much of 2003 and 2004. While there were important flashes of military activity, in the main the group simply chose to take cover and to wait out Plan Colombia. With PC committed for a limited duration, with prospects of additional U.S. funding for Colombia threatened by mounting U.S. debt and a strategic focus on Iraq and Afghanistan, and with what the FARC had falsely hoped would be the end of the Uribe government in 2006, the rebels sat tight until 2005. At that juncture, they emerged from retreat with a series of devastating attacks on the Colombian military. Between January and June 2005, the FARC killed 300 soldiers in a series of well-orchestrated strikes. An attack by 500 FARC troops in June 2005 on a military base guarding an oil installation in Putumayo raised some serious questions regarding the surveillance and intelligence capacity of the Colombian government, as we observed earlier. All of this was designed to demonstrate, at a time when Uribe was struggling for a reelection in 2006, that the Colombian government had not at all succeeded in its attempt to reduce the military power of the FARC. With Uribe in office again, the FARC continued their sporadic and devastating attacks throughout the latter half of 2006 and into 2007.

Overall, PC has indeed debilitated the FARC. The FARC has lost the limited international support it had attracted when it presided over the *zona,* its mobility has been constrained, and its relations to arms, drugs, and other contraband have grown more complicated due to PC. In many respects, its military capacity is probably close to what it was when PC began. But power is a relative term. PC has attempted to weaken the FARC by weakening the group's ties to drugs and by strengthening the Colombian army through Plan Patriota and other initiatives. While the effect of the drug war on the FARC is unclear, what is certain is that the Colombian military has grown far stronger since PC's implementation. The armed forces grew from 280,000 in 2002 to over 380,000 in 2005. They are better trained and equipped, quicker and more mobile, and more specialized in combat activity. The FARC also has faced stronger contestants in the form of paramilitary forces, as we shall see shortly. Meanwhile, some progovernment estimates suggest that the ranks of the FARC have dropped to about 12,000 in 2006 from over 20,000. These figures cannot be substantiated and are probably politically motivated to support Uribe during an election year. The FARC's troop levels remain unclear. Despite their relative loss of power, the rebels are still rich and powerful enough to sustain themselves into the indefinite future and to cause considerable damage to the Colombian armed forces and society.

The FARC's staying power has been a reflection of its capacity to harness in its favor certain aspects of the RMA, just as its weaknesses are largely due to its failure to embrace other elements of the RMA. Let us consider some of these. First, the FARC has worked well within globalization and transnational capitalism to enrich itself to the point that the group may be undefeatable. Second, the FARC has made the most of the element of asymmetry and has proven to be quite flexible. Plan Patriota, for example, has catalyzed the formation of smaller units of the FARC to avoid detection. The rebels also have employed to their favor the tactic of strategic retreat. Through its mastery of strategic space, the FARC has avoided the multifaceted surveillance system of PC by hiding in cloudy areas or dense jungle, where satellite surveillance is limited. Further, the group possesses a sophisticated surveillance system of its own, as demonstrated by its ability both to avoid the U.S. and Colombian military forces and to launch successful attacks against them, especially in 2005 and 2006. It has also developed considerable communication and transit systems that have supported its military and economic affairs.

A severe weakness of the FARC has been its inability to attract widespread popular support and its related lack of ideological development and credibility. The FARC does not reflect a revolutionary capacity. Ideologically, the FARC has offered nothing original. It has echoed the views of other leftists in Latin America, while clinging to its original platform aimed at its traditional peasant base. For example, it has embraced agrarian reform. The FARC has also demanded a forgiveness of peasant debt, the fortification of national consumption of agricultural products, the establishment of alternative development projects (to coca cultivation), the redistribution of wealth, and so on.[97] It has opposed neoliberal economics, privatization, and free trade agreements such as ALCA and Plan Puebla Panama (PPP).[98] The rebels have criticized TNCs that hire paramilitaries to protect their exploitative ventures.[99] The FARC has constantly criticized what it deems to be a "terrorist" government that serves the interests of the rich. It has argued that "[c]apitalism in its essence and its political, economic and cultural practices, produces states administrated by fascists, terrorists, mafiosas, mercenaries and authoritarians to guarantee and defend their class interests."[100] Further, the FARC in 2006 bitterly criticized Uribe's attempts to "legalize" the paramilitaries by providing them with total "impunity" through the demobilization process.[101]

The problem, then, is not that the FARC lacks ideas. Rather, the crux is that the FARC has emphasized military procurement over ideological and political development. This has meant an ideological platform that is devoid of creativity and of a capacity for popular engagement. Ironically, perhaps the most significant threat to the FARC comes not from the right-wing trinity of Uribe, the paramilitaries, and PC, but from the peaceful Left. The Polo Democrático Alternativo, a new party that represents the country's democratic Left, won over 22 percent of votes in the 2006 presidential election and thereby emerged as the country's official opposition party. The Polo Party's strong showing marks the first time the Left has won so many votes—for example, the Union Patriota received just 6 percent of federal votes.[102] Rather than the old dichotomy between the Liberals and Conservatives in Colombia, which has been dominant since the 1830s, the country's official party politics during 2006–2010 represented an historic feat by featuring a debate between the Right and the democratic Left. This is a truly revolutionary development, and one that is likely to isolate the FARC even further. That is, the Polo has received far more popular support than the FARC precisely because it has harnessed the power of ideas rather than rely on criminally subsidized brute force. Beyond being eclipsed by the democratic Left, tens of thousands of Colombians took to the streets in July 2007 to protest the Farc's horrendous policy of kidnapping and its use of hostages as political pawns. This signaled, once again, that the group has grown seriously out of touch with the country's majority population.

Finally, it is worth noting that the FARC did not launch a campaign of violence to disrupt the 2006 presidential election as it had in 2002. It had already displayed its military might by early 2006 and did not pursue a violent trajectory during the election period for a number of reasons. Given the wave of democratic elections that have introduced leftist forces throughout South America, in addition to the updraft enjoyed by Colombia's social democratic Polo Party, the FARC would have isolated itself even further had it engaged in a re-run of its rampage during the 2002 election. It may also be that the FARC toned down its behavior in an attempt to leave open a door for negotiation with the Uribe administration. But any negotiated settlement between the FARC and the government would be a challenging affair, given the hard-line Marxist policies of the guerrillas versus the solidly right-wing penchant of the Colombian government. Further, the deteriorated strategic position of the FARC was demonstrated in March 2007 when the group indicated that it would have to hold its ninth conference by Internet to avoid any potential capture of its leaders. Overall, while retaining much of its military capacity especially in the realm of hit-and-run guerrilla operations, the

FARC has become increasingly irrelevant politically both within Colombia and within Latin America. The FARC has morphed into a ferocious and criminal armed group occupying the shell of what was once a legitimate leftist guerrilla movement.

Plan Colombia, the RMA, and Armed Rebels: The ELN

Let us begin with a consideration of the ELN's ideological stance during the era of PC. The group's positions have not changed much since its inception. In addition to redistribution of wealth and agrarian reform, it continued to demand that TNCs leave more profit behind, raise wages of Colombian workers, and place more Colombians in managerial and other high-level positions. It has updated these traditional demands by opposing the ALCA free trade initiative and by protesting against the wave of privatization that reduces the power of unions. The ELN harshly criticized newfound U.S. oil interests in Colombia since 9/11. The rebels also have rejected the policy of extradition adopted by the Uribe government.[103] Enshrining the Latin American nationalism of Bolívar, the ELN has embraced the new brand of nationalism epitomized by Argentina's Nestor Kirchner and Venezuela's Hugo Chávez.[104]

Given that the ELN has a record of being far less murderous than either the FARC or the AUC, and has ideological views similar to those who have garnered widespread democratic support such as Venezuela's Hugo Chávez and Bolivia's Evo Morales, the question arises as to why it has not been able to attract more local and global support. Part of the answer lies with the ELN's relation to the off-putting web of crime-terror-war. Besides extortion, the ELN's penchant for kidnapping, or "retentions," has rendered its members as criminals in the minds of many. The group is responsible for about 25 percent of the 13,616 kidnappings reported during 1996–2003, and since then has continued in this endeavor but at a slower pace.[105] Some of these episodes have received considerable media attention as the ELN had hoped, as when the group kidnapped six foreigners hiking in the Sierra Madre in 2003. When the ordeal was finished, the rebels stated that "[t]he successful culmination of the liberation of foreigners demonstrated, once again, that the ELN is moving away from and rejecting terrorism."[106] But many observers viewed the ELN's kidnapping extravaganzas as a form of terrorism, and had a similar assessment of the group's destruction of energy towers, oil pipelines, and so on. As with the FARC, the ELN's participation in crime and perceived terrorist activities has served to neutralize the power that might otherwise have been derived from the group's ideological positions.

The ELN has suffered the strongest blows from PC and has been the least adept among the country's belligerents at managing the RMA. In the 1980s and especially into the 1990s, the ELN's strong suit was its ability to "tax" or extort money from TNCs operating in Colombia's extractive sector. In a very limited sense, the ELN was successful at tapping into the element of globalization associated with the RMA. But that source of income withered during the PC era, when the United States bolstered security along petroleum pipelines, when TNCs increasingly hired sophisticated private security corporations for protection, and when more reports emerged of paramilitaries guarding private enterprise.

The ELN has done a particularly poor job at being flexible in the face of a changing strategic climate. In Colombia's shark-infested military waters, the ELN has succumbed militarily to the FARC, to Colombian-U.S. military forces, and especially to the AUC. By early 2003, the group's numbers dwindled from an estimated peak of about 5,000, when PC commenced, to about 1,500–2,000. While these numbers may not be precise, for reasons mentioned earlier, the general trend toward a smaller rebel force has been apparent through 2007. Because the ELN comprised many nonprofessional soldiers who had day jobs or who worked during certain months of the year, the group has had limited mobility, has been less adaptable to strategic space, and therefore has been more vulnerable than its ferocious competitors.[107]

The ELN had attempted to cope with its steep decline by aligning with the FARC in the early part of the new millennium. The ELN has officially indicated that "[i]n diverse regions of the country we coordinate military operations (with the FARC). Both organizations are autonomous and sovereign."[108] But this has been an uneasy marriage. For decades, the ELN and the FARC have competed militarily and ideologically in Colombia, often in the most scathing manner imaginable. More recently, during the era of PC, the ELN has ceded power to the FARC in key strategic areas such as Arauca. Desperation at the prospect of disappearing altogether prompted the group's alignment with the FARC. But this arrangement seemed to fall apart by late 2006 and early 2007, when the ELN seemed to be more serious regarding peace negotiations it held with the Colombian government in Havana. Indeed, that development seemed to prompt the FARC to pursue frontal military attacks on the ELN, especially in strategic territory near the Venezuelan border. From Havana, the ELN leadership in February 2007 implored the FARC to halt its military offensive against the group.

Let us step back a bit, and consider the context of the ELN's negotiations with the Colombian government. The ELN in 2004 entered into what promised to be a thorny process of negotiation with the Uribe administration. The rebels in April 2005 temporarily rejected Mexico's participation in the process due to its criticism of Cuban human rights. A more positive sign appeared in September 2005, when a key ELN figure, Francisco Galan, was released from prison to engage in negotiations. Prospects for peace between the government and the ELN faced formidable challenges. Ideologically, the hard-core leftist demands of the ELN were highly unlikely to be met by the right-wing government of Uribe. Further, with reference to the AUC demobilization process, the ELN in May 2006 argued that "to have the conditions for peace in our country an essential factor is that there be no impunity, and above all, crime cannot be a method of political struggle."[109] In December 2006, the government pointed out that the ELN had kidnapped at least 54 people since negotiations began in 2004. Given that the ELN is not likely to achieve an agreement with the government on social policy issues, the only thing the rebels have left to win through negotiation is their own impunity from charges of subversion and crime. Finally, the government made it clear in October 2006 that if negotiations with the ELN are indeed successful, there will be no monetary compensation for demobilized leftist rebels as there has been for their right-wing counterparts, the AUC.

Plan Colombia, the RMA, and Armed Rebels: The AUC

> The first law of Colombia is impunity and our first unpunished offender is the President.[110]
>
> *Fernando Vallejo,* Our Lady of the Assassins

Of any single actor in the Colombian imbroglio, the AUC has enjoyed the most success with PC—although flies in the ointment appeared in late 2006. This umbrella group of paramilitaries, many with clear links to narcotrafficking, has shared almost identical strategic goals with the U.S. and Colombian governments vis-à-vis other armed groups. The AUC also has had mutual interests with them regarding issues of political economy, such as privatization, TNC investment, and so on. Importantly, it has viewed itself as a substitute for a state too feeble to function adequately in many parts of the country. Thus, the AUC has stated that it does not "demand the destruction or transformation of political, economic and social structures of the State and society," as have the FARC and the ELN.[111] Instead, the AUC has craved legitimacy within the state. Beyond representing the interests of drug traffickers and the rich, they have also represented small- to medium-sized landowners, those who resent the incursions of the guerrillas, and those who want more order in a chaotic society. Thus, the local support base for the AUC has been significant and nuanced.

The AUC has worked well with certain elements of the RMA. Being a highly flexible and expanding force that has gained the brute respect of leftist guerrillas as well as that of the U.S. and Colombian governments, the AUC has made the most of asymmetry. It has also mastered elements of illicit globalization. Even more than the FARC, the paramilitaries have been the epitome of the horrific trinity comprising crime, terror, and war. U.S. officials estimate that the AUC controlled about 40 percent of Colombia's cocaine exports.[112] They have been the leading human rights violators during much of the war, responsible for as much as 80 percent of human rights violations during key years. Of interest is the shift in leadership from Carlos Castaño, who was allegedly murdered at the hands of rival AUC forces in April 2004 but who is rumored to be alive, to Salvatore Mancuso. This represented a changing of the guards from a small faction of AUC members represented by Castaño, who were not directly involved in the drug trade or who wished to leave it, to what one Colombian security specialist called "30 Pablo Escobars" under the leadership of Mancuso.[113]

While the AUC's participation in criminal deeds is clear enough, its use of terror also is renowned.[114] It has murdered thousands of peasants during its tenure, including 1,895 assassinations from December 2002 to September 2004, which includes the early phase of its peace negotiations with the government and the period leading up to it.[115] Colombian analysts have referred to the AUC as *empresarios de la coercion* (businesspeople of force).[116] The AUC has muscled into urban areas using brutal violence and probable collusion with the Colombian armed forces, as it has in Medellín in 2002 and Arauca since 2001, and through the murder of 800 people in the oil hub of Barrancabermeja between 1999 and 2001.

The AUC has manipulated successfully other aspects of the RMA, including those associated with political economy, time, and space. Its participation in narcotrafficking already has been mentioned, with at least 18 paramilitary commanders populating Washington's list of top traffickers in 2005. Besides their successful participation in the transnational business of illicit drugs, paramilitary forces have controlled many local economies. They dominated contraband sales of gasoline, rice, cars, textiles, and other items. They have been involved in schemes regarding palm oil plantations that have taken over 21,000 hectares of collective community land away from blacks in Chocó.[117] Their skill at navigating the underworld political economy is unmatched. Their sense of strategic space has been keen, and the paramilitaries have demonstrated high mobility and flexibility. Their timing has also been stellar; they rose in power throughout the 1990s during a power vacuum and gained popular strength whenever it appeared that the FARC or the ELN might be on the ascendant.

The AUC has embarked on a U.S.-sponsored process of legitimization aimed at its reincorporation into society. The United States has donated in 2004 and 2005 relatively small amounts to aid the AUC's demobilization process, with a possible $160 million waiting in the wings—or about $8,000 per AUC member. Supporters of the process say it is a step forward to ending the armed conflict in the country, because one of Colombia's most notorious rebel groups would be legitimized. There are other perks. The AUC has bragged that it could be helpful to the U.S. and Colombian governments as it specializes in military techniques associated with "nonconventional wars," and that it looks forward to "inclusion" into the Colombian armed forces and public sector.[118]

Reminiscent of the FARC's *zona*, the AUC received in 2004 a 142-square-mile negotiation zone at Santa fe de Ralito, in addition to other areas. This was to be the central site for a demobilization process wherein the *paras* put down their arms to begin new lives as legitimate members of Colombian civil society. While there were estimated to be approximately 20,000 AUC members by 2006, considerable alarm was raised when the government announced that between 2003 and March 2006 some 31,000 AUC members were demobilized, which included 27,000 who were collectively demobilized and 4,000 individual cases. These inflated numbers are highly suspicious, to say the least, and strongly suggest that the process may have involved a shell game. That is, there may have been a cycle in place whereby thousands of paramilitaries disarm, but to some extent are replaced by others. In fact, new groups of *paras* have made their presence known in Nariño, Valle de Cauca, Antioquia, Chocó, Cordoba, and elsewhere.[119]

The demobilization process has been underpinned by the Law of Justice and Peace, signed into law by President Uribe in July 2005 and the subject of almost constant constitutional and legal haggling ever since. The law is another example of the paramilitaries' unquestionable influence over the Uribe government and over the Colombian Congress. Due to their warm relationship with the president, and their own declaration that they control more than one-third of Congress, it is clear that the paramilitaries represent an important pillar of the Colombian state. Further, paramilitary demobilization has had the strong backing of the Bush administration, and it was announced in September 2005 that Bill Clinton, a key architect of PC, would join the Friends of the Demobilization Process.[120]

But the law has evoked harsh attacks from U.S. politicians and global human rights NGOs. Amnesty International has suggested that the Law of Justice and

Peace essentially is a gift to the AUC that provides immunity to Colombia's single biggest perpetrator of crimes against humanity.[121] The law prevents the extradition of paramilitary members, and generally provides maximum criminal sentences of five to eight years, 18 months of which can be served by their time at the Santa fe de Ralito zone. The *New York Times* has deemed the law to be the "Impunity for Mass Murderers, Terrorists and Major Cocaine Traffickers Law."[122] Further, the International Commission of Jurists, among a chorus of other groups, has voiced loud complaints regarding the "lack of a true mechanism of verification" for the demobilization process. It also expressed doubts that the AUC will cease in its quest for the "control of cultivable territory."[123]

The demobilization process hit the bricks in a process beginning in late 2006. In the fall of that year, it became clear that the AUC had established strong links throughout the Uribe government—even beyond the large portion of Congress that the group has claimed it controls. By November of 2006, it seemed that key Colombian politicos were tied to the AUC on an almost daily basis. For example, the ex-director of the DAS, the country's intelligence agency, was exposed in November 2006 of having strong links to paramilitary forces—a charge for which he was arrested in February 2007. The influence of the *paras* proved to be increasingly pervasive in the country, with evidence in December 2006 of their extensive infiltration of the country's beloved game of soccer. The paramilitaries were also hand-in-pocket with key U.S.-based corporations. In March 2007 Chiquita Brands International was fined $25 million in a U.S. federal court when it admitted making payments totaling $1.7 million to paramilitary forces between 1997 and 2004.[124] Also in March 2007, a civil lawsuit gained momentum in Alabama, which alleged that Drummond Company's coal operations in Colombia facilitated the assassination of three union leaders at a mine in the northern part of the country.[125]

All of this contributed to an erosion of legitimacy for the Uribe government. Desperate politically, Uribe in late 2006 reacted by jailing paramilitary leaders in an attempt to show some political distance from them. By so doing, he clumsily converted the peace process into a criminal affair, leaving the AUC leaders with little reason to continue negotiation. By December 2006, the AUC officially withdrew from the peace process, leaving its future uncertain. The situation worsened in 2007, when over 60 government officials were cited by the Supreme Court as possibly having links to the *paras*. Among these were ex-president Horacio Serpa, the ex-minister of mines, the father and brother of the minister of foreign affairs Maria Consuelo Araujo (who was forced to resign), and a long list of others, including governors, members of Congress, mayors, the ex-director of the anti–drug use program, and military leaders.[126] Also in early 2007, ex-paramilitary chiefs indicated that demobilized *paras* were in a process of rearming at an alarming rate.[127] This crisis, dubbed by locals to be the "parapolitica," clearly exposed the extensive links established between the paramilitaries and various wings of the Colombian government, thereby reducing the legitimacy of the Uribe government, the demobilization process, and PC.

Despite all of that, there have been flashes of positive effects emanating from the demobilization process, at least in the short run. In order for the *paras* to achieve the essential impunity that the demobilization process yields, they have had to decrease, though not eliminate, their penchant for extraordinary violence. The paramilitaries have mellowed militarily during the process, in an obvious attempt to demonstrate to Colombians and the world that their legitimization will provide clear and fruitful results. We observed earlier that human rights abuses, and violence in general, have decreased since the demobilization process commenced. Ironically, this has also provided political space for the democratic Left in Colombia to a point never before witnessed in the country. That is, it was observed that the Polo Party gained over 22 percent of the votes in the May 2006 presidential elections—a watershed that was facilitated in part by the paramilitaries' relatively "good" behavior. Unlike the experience of the Union Patriotica during the late 1980s and early 1990s, Polo members have not yet been exterminated by the thousands at the hands of paramilitaries. If the trend continues, this could also bode well for the development of civil society in Colombia by diminishing the level of terror and exclusion in the country.

But disturbing questions remain. Is impunity for the horrendous crimes of the AUC a price worth paying for less violence and for the establishment of political space by the democratic Left and by critical elements of civil society? Did the "parapolitica" crisis that emerged in late 2006 and continued into 2007 mark the end of paramilitary negotiations, and will this mean they will rearm beyond the regrouping that has already occurred? Given their clear ties to multimillion crime, does the tight relationship between the paramilitaries and the state entrench a "narcogovernment"? Has the wobbly demobilization process simply been a shady scheme to construct a more formidable Colombian state that entrenches the interests of the country's illicit elite together with those of transnational capital?

Plan Colombia, the RMA, and Social Forces: A Revolution for Whom?

Given that PC is a U.S. initiative, let us begin by considering its effects from the perspective of Washington. The policy of aerial fumigation, and the drug war in general, have been dismal failures. But there have been some favorable short-term results. The United States has succeeded in diminishing the relative power of the FARC. American economic interests have been projected through a proposed free trade agreement with Colombia and through increasing investment especially in the petroleum sector. Security issues associated with them have been addressed through PC's military programs. For example, PC has provided substantially better protection of oil installations, while private security corporations and paramilitaries provided bolstered protection for an assortment of foreign corporations involved principally in the extractive sector. PC has strengthened—at least until the 2006–2007 scandal linking the government to the *paras*—the political fortune of Uribe, Washington's closest ally in South America. The United States has also succeeded in rendering Colombia as a bastion of U.S. military and political support in the strategic Andean region where anti-Americanism has run high. Indicative of this was Bush's seven-hour trip to Bogotá in March 2007, the first time a U.S. president visited Colombia's capital since 1982. Bush and his entourage heaped praise on PC and the Uribe government, and promised to try to secure an additional $3.9 billion in U.S. military assistance through 2014.

The Uribe government clearly has benefited from the military support provided by PC, and this contributed to his victory in 2006. The plan has helped Uribe to instill a greater sense of law and order in Colombia than had existed during previous administrations. Police and military forces increased in number, and were better trained and armed. Homicides dropped from 29,000 in Uribe's inaugural year of 2002 to 17,726 in 2006, while reported kidnappings fell from 3,000 to 800 between 2002 and 2005.[128] Basic travel along the country's highway systems became safer, permitting more circulation than had been possible for years. Further, conditions of social order were sufficiently established for the social democratic Polo Party to enjoy tremendous initial success during the 2006 election, instilling a hope for real democratic progress in the country. With regard to economic issues, we observed that GDP growth rates advanced considerably during the first term of the Uribe government as did foreign investment, with official unemployment rates falling to about 12 percent from around 20 percent at the beginning of the decade. Thus, Álvaro Uribe has not only represented the interests of the United States and transnational capital, but has also succeeded in providing a greater sense of law and order as well as more jobs to ordinary Colombians. That is why he received almost two-thirds of the votes cast by the 45 percent of eligible Colombians who showed up at the polls in May 2006.

But despite these clear achievements, a bleak laundry list of problems has persisted. Colombia's horrific poverty and inequity rates remained in a general holding pattern. While jobs were created, these overwhelmingly tended to be lower-paying, less secure, nonunionized positions. Unemployment, though lower in 2006 than it had been for years, remained at about 12 percent, while subemployment hovered at about 36 percent. As we observed, inequity in Colombia during the Uribe administration was among the worst in the world. Labor organizations have been under siege during his government, and the displacement continues to fester through 2007. There has been vigorous global criticism of the law that framed the demobilization process for the *paras*. Further, in an unusually scathing critique, the United Nation's agency that focuses on human development criticized the Colombian government for the huge levels of corruption, clientelism, nepotism, criminality, and extreme political exclusion in the country.[129] Regarding the issue of political support, it has been emphasized that less than half of the country's eligible voters cast a vote in the 2006 presidential election, providing a sobering backdrop to Uribe's "landslide" victory. Finally, the legitimacy of the Uribe government has been broadsided with the emergence in late 2006 of the "parapolitica" scandal—one that suggested an axis between the government, the military, intelligence agencies, U.S. corporations, and narcoterrorists in the form of Colombia's thriving paramilitary forces.

It will be recalled that Alberto Fujimori initially was wildly popular in Peru for many of the same reasons that explain Uribe's popularity. The Peruvian experience demonstrated that when people live under conditions of war-torn chaos, as Peru did under the siege of Sendero Luminoso, what they crave first and foremost is social order. Once that is established, they long for a real political voice, and for more equity, higher wages, and social programs aimed at the majority population who are poor. That is where Fujimori failed in the long run. Uribe's second term will demonstrate if he has learned from Peru's experience.

* * *

Historically, civil society has been quite weak in Colombia. The director of one of the country's leading NGOs indicates that it has been extraordinarily difficult for civil society to develop in a climate of war, terror, and frequent assassination of civil rights leaders.[130] Further, civil society has remained excluded from negotiations regarding the armed conflict, such as those surrounding the demobilization of paramilitary groups. And like the rest of politics in Colombia, civil society typically has been fragmented. It has not yet gained the strength and solidarity to attract widespread global support. That is, part of the RMA entails the exploration of new political organizations, connectivity, and dissemination of key information. But Colombian civil society has not yet conjoined in any widespread fashion with global social movements, and it has not yet organized in a manner that has been successful in attracting sustained and potent global attention. We shall revisit this general issue in Chapter 7 when we explore the relative success of Mexican civil society in attracting the global gaze.

What is clear is that PC, as an expression of the RMA, is a revolution from above. It has been used principally to advance U.S. strategic and economic interests. The short-term results have been positive for many Colombians, but this progress has been highly uneven and has been absent for many. The two social forces considered here, organized labor and the displaced population, in many ways have witnessed increasing hardship during the PC era. In a country with astounding inequity and poverty, the swan song of organized labor cannot be viewed as a positive development. Unions have shrunk through neoliberal policies of privatization, and this has been twinned with brutal repression largely on the part of paramilitary forces that have been fortified during PC. Further, the displaced population has continued to swell considerably during the plan's tenure, violently pushed aside by power-maximizing belligerents.

While the belligerents possess only two features of Thucydides' trinity of fear-interest-honor, civil society in Colombia has all three. It has waged a courageous and honorable struggle for basic human rights, political inclusion, and escape from economic and other exploitation. Members of labor organizations, for example, risk their lives on a daily basis struggling for decent wages and working conditions. The hemisphere's largest displaced population has become social orphans who struggle just to survive in hostile environs. The strongest ray of hope for Colombian civil society since the 1980s has been the enormous success of the Polo Party in 2006, as it raises the specter that NGOs and other components of civil society may also blossom for the first time. Also demonstrative of the growing power of civil society was the historic march thought the country's major cities in July 2007 to protest against the FARC's brutal tactic of political kidnapping, an event that drew tens of thousands of people. Ordinary Colombians are tired of playing victim to a host of self-serving belligerents.

We observed the irony that the fortification of Colombia's still weak civil society has been facilitated in part by the subdued behavior on the part of paramilitary forces during a crucial point in their attempt at legitimization. If this trend endures, it could open the door for the development of the country's civil society as never before. But with the demobilization process in crisis in 2007, and with no incentive for long-term pacific behavior in place, it is not clear if the paramilitaries' relatively good behavior will last. If it does not, the country could implode once again and revisit the violent depths of the post–Union Patriotica period.

Strategic change in Colombia since the turn of the new century has been linked to the RMA. In the previous chapter it was observed that various features of the RMA worked together synergetically for a powerful and exponential effect. Components such as the discourse on terror, asymmetric warfare, complexity, and surveillance have been conjoined through PC. This has been reflected in the vast restructuring of the Colombian army both temporally and spatially, the utility of an array of both high- and low-tech surveillance devices to limit the movement of leftist subversives, the demonization of leftist guerrillas through their new portrayal as terrorists, and so on. As part of this larger system of systems, privatization also is a key theme of the RMA. In this chapter we explored its ties to relatively newfangled elements of the Colombian imbroglio, such as the use of private security and military corporations, the use of neomercenaries, and so on. Privatization has been a fundamental aspect of refashioned asymmetric warfare, and has relied considerably on aforementioned ultra-surveillance. And we observed the trajectory since the 1980s of an escalation and caricaturization of warfare in Colombia that moved in step with the private financing of insurgents through the illicit transnational commerce of narcotrafficking. Against the backdrop of the failed U.S. invasion of Iraq and the increasing preciousness of Colombian oil for Washington, it was observed that the creeping privatization of the extractive sector has also been a growing source of armed conflict.

Overall, the point is that many features of the RMA conspire to explain strategic change over the last decade or so. The U.S. and Colombian governments, and their paramilitary allies, have fortified their short-term positions to the extent that they have successfully exploited an assortment of features associated with the RMA. But PC will likely fail overall due to a myopic strategic vision—one that concentrates on force and the interests of TNCs rather than on the well-being of the majority population. Turning to civil society in Colombia, we have observed that it has not yet utilized in a sweeping way elements of the RMA to advance its interests. But there indeed have been flashes of success for civil society. Noteworthy in this regard is the case of organized labor, which has benefited from fewer human rights abuses since the establishment of international surveillance over Colombia's abysmal environment for union members. With an ILO office open in Bogotá, and better links established between Colombian labor and their global counterparts, surveillance as an element of the RMA has furthered the interests of labor. Through such surveillance, labor organizations have been exploring the establishment of global networks that are key to the RMA. But this is a work in progress, with a long road ahead. We shall return to an analytical discussion of the RMA and its relation to strategic change in Colombia in the concluding chapter. Let us turn now to a consideration of the Mexican case. We shall see that many of the same features of the RMA expressed in Colombia during the PC era have also been apparent in Mexico, but with distinct manifestations. If PC represents an RMA from above, we shall see that Mexican civil society has witnessed mixed success by forging an RMA from below.

5

Historical Aspects of Mexico's Strategic Landscape

Comparing Colombia and Mexico—Some Initial Considerations

The last three chapters focused on the recent manifestation of the RMA in Colombia, especially since the commencement in 2000 of the multibillion dollar military extravaganza under the banner of PC. Features of the RMA have been filtered through the unique circumstances of the Colombian imbroglio combined with pronounced U.S. strategic interests in the region. The RMA is never expressed identically in any two countries due to particularities associated with local politics and global geopolitics. Yet certain cases can yield useful comparisons, given shared features of the RMA and the presence of concentric areas. Mexico and Colombia, for example, are among the very few Latin American countries to host significant guerrilla movements in the new millennium, albeit they are very different in nature and context. They are also the only two major governments in Latin America, which, by 2007, bucked the regional trend and continued to embrace U.S.-styled neoliberalism. Related to this, they have been Washington's two closest strategic allies in the region. Further, both nations have witnessed social conflict emanating in part from debt and inequity. Along with Brazil, they represent about half of Latin America's poverty-stricken population.[1] And they both have endured the intense strategic effects of narcotrafficking and other types of crime to the point that some refer to the "Colombianization" of Mexico.

With regard to other initial points of comparison, the relation between war and terror was highlighted in the Colombian situation. In Mexico, the specter of terror has been far less intense than in Colombia. This is largely due to the once-formidable hegemony of the Mexican state and an enduring record of general political stability in the country, and is also a result of a political culture less tolerant of violence than is the case in Colombia. Yet important flashes of terror have indeed been present in Mexico, both historically and recently, especially on the country's periphery where hegemony wears thin. There have been instances of the use of terror by insurgents, the state, and, in its most extreme and recent form, criminal syndicates. Further, the discourse on terror has been manipulated by the state in relation to insurgent groups. Asymmetry, a related and important feature of the RMA, is also apparent in both countries. Asymmetry is a phenomenon faced not only by the state with regard to guerrilla forces but, especially important for our purposes, by components of civil society in their struggle for social justice vis-à-vis the state and transnational forces.

Complexity is a vital feature of the RMA, as the Colombian case so clearly demonstrates with regard to the implications of almost incessant warfare since the country's creation, the escalating and corrosive influence of narcotrafficking and other kinds of mega-profitable crime on the part of both insurgents and the state, the lack of discernable heroes in the violent struggle there, the ecological and social disasters produced by the failed policy of fumigation, and so on. With regard to Mexico, complexity is apparent especially with respect to the contradictory forces of economic integration under Nafta versus the post-9/11 Fortress America connotation of a new "iron curtain" at the U.S.-Mexican border. Complexity as a reflection of the RMA is also exemplified with the legendary use of info-war and netwar by civil society and insurgent forces in Mexico. Here there is a link to other features of the RMA, namely, surveillance and intelligence. We shall also investigate their use with regard to the crucial issue of migration.

Finally, there are important commonalities between Colombia and Mexico with regard to the privatization of warfare and the redefinition of political space. While Colombia is unique with regard to the use of neomercenaries within PC, both Mexico and Colombia have hosted privatized war in relation to transnational crime. Also, both have faced intense social struggle based on plans to privatize former state property in the name of neoliberal restructuring. A crucial phenomenon in the Mexican case is the reclamation of strategic space by social forces, as evidenced by the Caracoles in Chiapas and the PCs in Guerrero. Overall, Mexico and Colombia represent good comparative cases for the manifestation of the RMA in Latin America, due to their broad range of similarities and distinctions.

Most importantly, it is the synergetic cocktail of factors associated with the RMA that best explains strategic change in both countries. That is, what is significant is not any single feature linked to the RMA, but how these elements work together such that the strategic result is bigger than the sum of the individual parts. For example, neoliberal restructuring that entails government cutbacks of social programs weakens the legitimacy of the state, and this is amplified even further by corruption emanating from criminal factors such as narcotrafficking, which simultaneously empowers state enemies. State power is also debilitated by the privatization of warfare, by the dispersion of information and weapons, and by the advent of identity politics that dissolves national patriotism into more fragmented political loyalties. This brief example focused on fading state power, but the synergetic effect of the RMA's various components can explain the changing strategic position of assorted social forces, insurgents, criminal syndicates, and so on. The point is that particular factors associated with the RMA conspire for an exponential strategic effect. This idea will be developed in the "Conclusion," where the Colombian and Mexican cases will be examined with regard to the synergetic influence of the RMA.

Beyond the typical assessment of the state and insurgents, this study also examines social forces in relation to the RMA. We examined the role of labor and the displaced as national phenomenon with regard to Colombia. When analyzing social forces in the Mexican case, the country's size and diversity need to be taken into account. This is not to suggest that Colombia lacks diversity or variation, but that these features are amplified even further in Mexico. Some of this has to do with Mexico's vast geography, occupying some 1,923,040 square kilometers compared to Colombia's 1,138,910. There is also Mexico's multifaceted predicament of having the United States as its next-door neighbor. Thus, there are many Mexicos—from the quarter of the country's population that resides in cosmopolitan Mexico City, to the Americanized border towns in the north, to the tourist havens for wealthy foreigners on the coasts, to the indigenous and relatively impoverished south, and so on.

Given the relative vastness and diversity of Mexico, one way to get a solid feel of the plight of Mexican social forces in relation to the RMA is to provide a national overview, and to complement this with a special focus on a microcosm of sorts through which the RMA filters. That is, by concentrating on social forces within a relatively small geographic space, social struggles and their meaning to civil society can be brought into sharper focus than by relying solely upon piecemeal examples from disparate parts of this vast country. Such an approach emphasizes a humanistic dimension to the notion of security. Guerrero will serve as an analytical centerpiece for the Mexican case for two main reasons. First, it is Mexico's most "Colombianized" state, so important parallels and distinctions between the two countries can come into focus through an examination of Guerrero. Second, although some aspects associated with the RMA, such as guerrilla insurgency, have been largely confined to pockets of southern Mexico, many of the manifestations of the RMA that affect Guerrero represent broader national phenomena. Indeed, Mexico's secretary of the interior, Francisco Ramírez Acuña, suggested in 2007 that Guerrero represents a "red light" for the security of the entire country, given the panorama of the state's alarming security problems.[2] While Guerrero makes a good comparative focal point, a wide assortment of other Mexican localities will also be discussed at length in relation to the RMA.

The analysis of Guerrero will make frequent reference to the cases of Chiapas and Oaxaca, which together represent the three poorest, the most inequitable, and the most volatile states in the country.[3] The United Nations Program on Development (UNDP) has deemed this trio to be the last on the list for human development in Mexico.[4] Chiapas is an important point of reference because the 1994 Zapatista rebellion has had an enormous influence on social struggle throughout the country, and because the newfangled aspects associated with the EZLN's strategy provide a sharp contrast from which to consider the more traditional approach in Guerrero. Oaxaca, the site of extreme social tension in 2006 and 2007, also is akin to Guerrero. Both states have been the primary sites for the EPR rebel group. They share similar levels of political organization but are distinguished by factors such as the relatively high indigenous population of Oaxaca and the peculiar strategic circumstances surrounding it.

Organization of the Chapters on Mexico

This chapter will begin with a brief discussion of the history of RMAs in Mexico as well as other contextual features of the current RMA in the country. The focus will then shift to the historical context of Guerrero and, especially, its legacy of state terror and asymmetric warfare. We shall highlight Guerrero's role as the primary site for the country's notorious Dirty War—the vivid ghost of which hovers weightily over the country since the release of an extraordinary government report in 2006 documenting horrific acts of what the government itself has deemed to be "terror" and "genocide."

Chapter 6 will commence with an examination of the contextual implications of the neoliberal era for Mexican security. The chapter features a look at the dual phenomena of fight and flight in relation to the RMA. With respect to "fights" or armed struggle, there will be an analysis of insurgency, asymmetry, and terror dating from the 1990s into the new millennium. With regard to "flight" or migration, the analytical features of complexity and surveillance will be discussed.

Chapter 7, the third and final chapter on Mexico, will explore the twin themes of privatized war and public resistance. The trinity of crime, privatization, and the erosion of security will be examined. This will be followed by a discussion of the emergence of largely peaceful social forces that are struggling to use elements of the RMA in their favor in an energetic attempt to construct what they view as a legitimate conception of human security. There will be an emphasis on identity politics in relation to social forces—including a discussion of social strife in Oaxaca, and the significance of Caracoles in Chiapas and of PCs in Guerrero, among other topics.

A History of Mexican RMAs

Before exploring aspects of Mexico's relation to the current RMA, it may be helpful to situate the RMA within the context of the country's experience with a series of previous ruptures. Certainly, the sixteenth-century Spanish Conquest is noteworthy in this regard. So, too, are the ruptures associated with Mexico's independence in the nineteenth century and its revolution from 1910 to 1920. The country's most recent watershed commenced in 1994, and was heralded by the creation of Nafta and the pivotal appearance of the postmodern EZLN in Chiapas.

To start, a brief comparison to the Colombian case will be helpful to flesh out the context from which to consider Mexican RMAs. A major historical distinction between Mexico and Colombia is that Mexico hosted two great civilizations prior to the Spanish conquest, the Mayan and Aztec empires. This was not only a unifying force during the heyday of the empires, but also served as a common point of reference after the conquest. That is, rather than the sense of fragmentation that continues to haunt Colombia, Mexico could draw upon a common social identity in its generally successful quest to centralize power within the confines of a modern nation-state. Let us consider more carefully the implications of a shared Mexican identity before and after the Conquest.

The Conquest

While what is now Mexico had its own unifying forces within the Aztec and Mayan empires, the Spanish attempted to impose a distinct and broader sense of unification under the Conquest. The vastness of that project entailed the imposition upon Latin America of a singular entity unified by the Spanish language. It was further characterized by a pseudo-feudal economic system, an authoritarian political structure, a culture of individualism, and Catholicism.[5] Through this process, the indigenous and their culture were devalued to the extreme.[6] Further, the indigenous population of central Mexico was devastated by disease, with a preconquest population of 25 million falling to 1 million in 1605. This meant sheer terror: the death of up to 95 percent of one's civilization, the flight of the gods, the imposition of an exploitative economic system under foreign rulers, and so on.

In a truly classic study of the country, Octavio Paz suggests that the Conquest resulted in a Mexican political culture characterized by the feeling of intense solitude combined with a sense of abandonment. This, Paz argues, is due to the belief at the time that the indigenous gods abandoned Mexicans in the face of far stronger deities—"The Conquest of Mexico would be inexplicable without the treachery of the gods, who denied their own people."[7] The effects of such a traumatic rupture linger within Mexico, Paz suggests. While one must certainly be cautious when totalizing the experience of all Mexicans, as there are indeed "many Mexicos" and myriad political identities, there nevertheless is validity to pointing toward the contours of a political culture that may exist at the national level, as Paz so brilliantly does. Most importantly for our purposes, this historical context sparked a noticeable tendency in Mexican society, once independence was achieved, to reestablish its identity through a mythical return to the pre-Conquest landscape of indigenous Mexico.[8] This theme was expressed not only by Paz but also by quintessential Mexican writers, such as Juan Rulfo in *Pedro Páramo,* whose classic work begins, "I came to Comala because I had been told that my father, a man named Pedro Páramo, lived there."[9] This attempt to return to origin was expressed, as we shall see later, in the Mexican Revolution and again among guerrilla groups and social movements since 1994.

Independence

In contrast to Colombia, which was liberated by the Venezuelan Simón Bolívar, it was the Mexicans themselves who liberated their country from Spanish colonial rule. Heroes of that epoch include Hidalgo, Morelos, and also Vicente Guerrero, who belonged to the area where a state was later named after him. Indeed, the state of Guerrero is renowned for producing some of Mexico's best soldiers, not only in the War of Independence, but also in the battle against the French as well as during the revolution—a point to which we shall return. What is important to emphasize here is that independence shed the Spanish yoke from Mexico, but Mexican Creoles—descendants of the Spanish born in Mexico—perpetuated virtually the same system as had the Conquistadors. The RMA associated with independence, then, was revolutionary only to the extent that it entailed a changing of the guards, since the essence of the previous system remained intact. The church grew even more powerful during this era, as did homegrown elements of the Mexican military. In fact, military strong men, or *caudillos,* together with the church set the basis for what was later known as the *cacique* system that linked the Partido Revolucionario Institucional (PRI), the church, the military and/or police, and local business.[10]

While the locals were potent enough to evict the Spanish during the independence era, in general Mexico remained debilitated and politically fragmented until the revolution of 1910–1920. This weakness did not go unnoticed by the United States. If Mexican independence meant more continuity than the nomenclature suggests, a true RMA was indeed fermenting in its northern neighbor, one that spelled unhappy consequences for Mexico. The United States in the nineteenth century was witnessing an astonishing ascendance with regard to both economic and military power, and in the 1823 Monroe Doctrine boldly signaled its intent to dominate the Americas. The United States voted in 1845 to annex Texas and proceeded to take one-half of Mexico's national territory by 1848. Given this context of U.S. encroachment, which transformed over the years from a military to an economic challenge, the United States was viewed as Mexico's single biggest external security threat from 1848 until the Salinas presidency of 1988–1994.

Hegemony and the Mexican Revolution

The revolution of 1910–1920 was Mexico's most important RMA between the conquest and the signing of Nafta. Unlike some RMAs to which we referred in the introductory chapter, this one was not about new military technology. Instead it concerned the myriad features associated with the enormous and largely successful jump from colonial feudalism to modernity. This entailed the formation of a modern nation-state, the forging of a nationalist political identity, and the increasing prevalence of notions such as progress, balance, secular politics, welfare, redistribution, and so on. With regard to the armed forces, the Mexican Revolution signaled the establishment of hegemonic consent that meant the military receded to the background of Mexican affairs amidst an atmosphere of stability that is second to none in Latin America. This represented a correctly balanced concoction of Machiavelli's "fox" and "lion."[11] Let us consider these features in greater detail.

The rupture was underpinned by a wide ideological and epistemological shift. The Revolution rejected the Spanish aristocracy and their Creole heirs, as well as the system of feudalism upon which it was based. Perhaps the most revolutionary aspect of this rupture was in terms of its answer to the question, a revolution for whom? This was a revolution that celebrated the interests of the peasants, the mestizos, and the Indians. In that sense, it was a return to Mexican origins, to its indigenous past. As Paz suggests, the revolution "affirmed that any political construction, if it is to be truly productive, must derive from the most ancient, stable and last part of our national being: the indigenous past."[12] This is crucial for our purposes, because this component of the revolution—the allegiance to the indigenous and to peasants—was betrayed in the latter part of the twentieth century.

Just prior to the revolution, 98 percent of the country's arable land was part of the feudal hacienda system, leaving about 90 percent of the population landless.[13] Zapata's piercing slogan of "land and liberty" embodied the essence of the revolution. Land reform was front and center, and was epitomized by Zapata's Plan de Ayala. The promise of this was realized most fully some two decades later during the Cárdenas presidency of 1934 to 1940. This approach established a bond between the new government and the majority population of Mexico, the peasants or *campesinos*. The state also embraced the interests of the small bourgeoisie and common workers.[14]

The proliferation of government-owned corporations, and the nationalization of the oil industry in 1938, exemplified the state's role as a principal agent of economic progress. Fueled by the nationalist import substitution model, Mexican economic growth rates averaged about 6 percent between the late 1930s and early 1960s. Free education, even at the university level, combined with policies of redistribution and land reform helped create the outlines of an emergent middle class by the 1970s. The Mexican Revolution, viewed as a protracted historical process and not just an event, provided a sense of unity, identity, and purpose.

This spirit of revolutionary nationalism, the ideological core of this epoch, not only yielded material progress but also provided the foundation for a remarkable record of political stability. The PRI—a party that was born from the revolution—was hegemonic until about the late 1950s, not only in the sense of its unchallenged dominance, but also because it was able to garner widespread consent such that the use of coercion was relatively infrequent. The RMA associated with this epoch, then, was not one built in any predominant fashion on brute force, although the capacity for that was present. Mexico did not host military governments characterized by a revolving door of coups and subversion as was witnessed by other Latin American states in nonhegemonic situations. In terms of external security interests, the worry was principally U.S. economic encroachment rather than the States' military adventurism of the previous century.

While hegemony rests on clear material benefits for the majority population, it was discussed in the "Introduction" that good intelligence services can help diminish the prospects for violence by identifying in advance issues that could emerge as serious security threats. The theme of intelligence will be central to the ensuing analysis. Mexico's first such institution was initiated in 1918, a year after the constitution was established. It assumed the official title of the Departmental Confidencial in 1929, just after the essence of what would become the PRI had formed. By the late 1940s, the agency engaged in classic intelligence tasks, complete with coded language, the interception of telephone calls, and the infiltration of counterconsensus groups.[15] This was the dawn of the cold war, with Mexican security apparatus now influenced by a world order that pitted U.S. capitalism against Soviet socialism. By now the Mexican government, with considerable U.S. pressure, found as its principal internal enemy elements of the marginalized population who sought social justice by embracing Marxism.

Within this context, the likeliest "suspects" were those who fell through the cracks of the system of benefits provided by revolutionary nationalism. These included certain labor organizations that became radicalized and wanted a more leftist political economy than expressed in Mexico's mixed system, and especially pockets of impoverished *campesinos* that did not reap the hoped-for benefits of the national land redistribution process. Land reform was an uneven process, temporally and spatially and also with regard to the quality of land distributed.

Shattered Hegemony and the Specter of a New RMA

The hegemony of the PRI began to shatter in a process that commenced in the 1960s and terminated during the Salinas sexenio of 1988–1994. This phenomenon marked the end of an era, and the dawn of a transformation toward another RMA in Mexico that materialized with the formation of Nafta. This "beginning of the end" phase was marked by state terror and what the Mexican government in 2006 deemed to be "genocide"—a point to which we shall return. Perhaps the opening salvo of this epoch was the notorious massacre of Mexico City university students who were protesting for greater national democracy on the eve of the 1968 Olympic Games. It signaled a turning point in Mexican history toward an era that would rely increasingly on force as consent wore thin. Also crucial in this regard was the appearance of guerrilla movements in Guerrero, Oaxaca, and other states in southern Mexico that were met with a Dirty War on the part of the government. Many hundreds of *campesinos* were murdered and "disappeared" at the hands of the Mexican military. Mounting economic problems appeared during this epoch. Not the least of these was an increasingly uncompetitive manufacturing sector during the twilight of the ISI. This was compounded by problems in the agricultural sector that included land shortages, increasingly limited access to government agricultural credits, and so on. The latter precipitated escalating social unrest in the countryside, especially in the impoverished southern region of the country.

The PRI's precipitous free fall hit full swing with the onset of the 1982 debt crisis. Revolutionary nationalism was trashed as neoliberalism was imposed upon Mexico through IMF restructuring policies. These opened the country to unlimited foreign investment in most realms, to decreased tariffs and taxes for corporations, and to drastic cuts in social welfare spending. Labor no longer received packages of PRI largesse, while *campesinos* watched government agricultural subsidies dwindle severely and thus cripple attempts at peasant subsistence. At the same time, the bourgeoisie fractured between the so-called dinosaurs of national capital and the quickly ascendant maidens of transnational capital. The foundation of the PRI crumbled, leaving it to rely increasingly on the use of force and fraud. The latter was most obviously demonstrated in the presidential election of 1988, where it is widely believed that Carlos Salinas actually lost the vote to his center-left rival, Cuauhtémoc Cárdenas.[16]

The debt crisis and the imposition of neoliberalism set the stage for a restructuring that underpinned Mexico's integration with the United States and Canada under Nafta. This meant the transition from national to regional security for Mexico, a phenomenon that prompted intensifying U.S. domination. Nafta represented one bookend for the onset of an RMA in Mexico. With its entrance into the trade bloc, Mexico traded the ideals of the revolution for the false promise of insertion into the first world.[17] Remarkably, and virtually overnight, the United States was recast from its previous role as Mexico's biggest external security threat to its ticket to top-tier economic status through North American integration.

The other bookend, as it were, was 9/11. The subsequent development of U.S. "homeland security" meant that Washington became increasingly concerned with Mexico's potential role as a conduit for "terrorists." Mexico's northern and southern borders, as well as its ports and airports, became a prime security preoccupation for the States. This RMA—situated within the context of a world order characterized by transnational capitalism and the U.S. fight against "terror"—transformed the roster of Mexico's enemies, redefined its strategic space and the meaning of its frontiers, and, more generally, altered in a radical way the country's discourse of security. Despite this shift, it is noteworthy that until 9/11, a celebrated Mexican analyst argued convincingly that crime represented Mexico's key strategic threat—especially in relation to narcotrafficking and the related illicit arms trade emanating from the United States. It is important to emphasize that this threat has continued to escalate. In other words, the post-9/11 world ushered in an array of new strategic threats for Mexico, compounding many of the existing threats in the country, which have amplified considerably.[18]

There are two additional global developments that have framed the manifestation of a fresh RMA in Mexico. The first of these has been the widespread Latin American rejection of U.S.-styled neoliberalism, as evidenced by the pronounced degree of polarization that at least until 2007 pitted an expanding wave of democratic leftism in Latin America against the two staunch supporters of U.S.-styled neoliberalism: Colombia and Mexico.[19] The socially harmful effects of this economic model have increasingly been rejected by the majority population of Latin America. While they were willing to accept the IMF argument of short-term pain for long-term gain back in the 1980s and early 1990s, that rationale no longer held weight after more than a generation of rising inequity, crime rates, and ecocide. While the governments of Colombia and Mexico have energetically defended the neoliberal model, increasing portions of their own population have suffered from its effects. This was especially evident with regard to the 2006 Mexican presidential election and its aftermath, when almost half of Mexican voters rejected the neoliberal model embraced by President Felipe Calderón.

Another contextual component of the recent RMA concerns a new constellation of global power whereby countries such as China are on the ascendant. The rising dynamo of China has been responsible for a considerable portion of the economic growth in Latin America since the beginning of the new millennium, especially in those countries that are exporters of petroleum, gas, and other natural resources. Chinese imports from Latin America grew about 600 percent from \$3 billion in 1999 to \$21.7 billion in 2004. By 2005, China imported 40 percent of its raw materials from Latin America, and about 40 percent of its global foreign direct investment was located in the region. China–Latin American trade reached

$50 billion in that year. In 2006, Chinese officials made numerous pronouncements that celebrated prospects of deepening economic ties with Latin America over the coming decades.[20]

In the big picture, there is no doubt that China remains dwarfed by the shadow of U.S. economic interests in Latin America. But China's rocketing power has provided the region with an alternative economic giant to which it can direct its aspirations, one spawned by a decidedly different model of development than advocated by Washington since the 1980s. It could be the beginning of glacially slow displacement of economic power in the region away from the United States. This became painfully clear to the Bush administration during the Latin American summit of November 2005, where the president was rebuked by most regional leaders as he repeated tired slogans in support of neoliberalism. Further, bled by the debt accrued from a disastrous war in Iraq, Bush arrived with empty pockets and the incapacity to buy political friends. A rerun of that phenomenon occurred during President Bush's 2007 trip to five Latin American states, at which time it was apparent that Venezuela's Hugo Chávez provided more developmental assistance to the region than the United States. In the global context, China's rising star came at a time when U.S. popularity and influence in Latin America reached new lows. Despite that generality, some in Latin America have viewed China as a threat of sorts. Certainly, this has been the case with respect to Mexico, which has seen its maquiladora sector decline precipitously against Chinese competitors.[21] While we shall address in subsequent chapters the security implications of declining employment, among other features related to the RMA, let us first consider the historical context of the social struggle in Guerrero.

A Fading Revolution: Insurgency, Asymmetry, State Terror, and Genocide

> We opened our mouths to say that we didn't want the plain, that we wanted what was by the river. . . . Not this tough cow's hide they call the plain.[22]
>
> *Juan Rulfo,* "They Gave Us the Land"

Let us proceed to examine the historical context of Mexico's crucial insurgencies in the state of Guerrero, which represented the epicenter of the country's notorious Dirty War. Remarkably, this horrific episode, which spanned from the late 1960s into the early 1980s, was never fully acknowledged by the country's authorities until the release in late 2006 of a pivotal government report. We shall examine this period in three phases, beginning with the prelude to the Dirty War and the emergence of legendary guerrilla leader Lucio Cabañas. The focus will then shift to the Dirty War itself, and finally to the subsequent epoch marked by the dawn of neoliberalism. All this will serve as a basis from which to analyze the expression of the RMA through more recent insurgencies in various parts of southern Mexico.

Prior to a consideration of the Dirty War, it is worth recalling briefly that, in the previous discussion of Colombia, the prominent roles of violence, fragmentation, and insurgency were emphasized. While the array of important distinctions between the Mexican and Colombian cases will be sorted out in the "Conclusion," here it is worth underscoring some similarities between them. Perhaps more than any other Mexican state, Guerrero resembles Colombia to the extent that historically it has been plagued with chronic violence.[23] In a similar fashion to the Colombian case, Mexican insurgencies generally have been the result of gross social inequity and political exclusion mixed with a fragmentation of power.[24] Further, Mexico's special prosecutor for social and political movements of the past, in an extraordinary government report made public in November of 2006, observed the historical presence of certain features in Guerrero that seem to echo those found in Colombia, especially the "frequent," "unnecessary," and "excessive" use of force, a recipe that has accompanied "pre-modern structures" of power.[25]

Phase One: Historical Context and the Emergence of Cabañas

Central to the political discord historically present in Guerrero and in much of southern Mexico has been the entrenched *cacique* system. Since the revolution and through the current juncture, it has meant a powerful nexus between the PRI, rich landowners, elements of the military and paramilitary forces (*pistoleros* or *guardias blancas*), and some leading members of the traditional Catholic Church. Typically, *caciques* maintain tight control over areas such as agricultural storage facilities, provision of seeds and fertilizers, product marketing, transportation, credit and banking, product sales, job provision, and so on.[26] Because of this extensive web of control, peasants may find themselves in servitude to *caciques* in what amounts to a feudal or premodern arrangement. Not only have *caciques* feuded among themselves through local power plays, but they have also encouraged bitter infighting between *campesinos* and between indigenous groups for land and other resources outside the *caciques*' immediate grasp.

Let us turn to some historical aspects of the *cacique* system in Guerrero. Given the prerevolutionary context of feudalism, it is not surprising that Zapata's stress on "land and liberty," and especially his Plan de Ayala, were wildly popular among the majority population of Guerrero's *campesinos*. They began to organize in the 1920s, with the formation of groups such as the Congreso Campesino de Guerrero (1923), the Liga de Resistencia Obrera y Campesina (1929), and the Liga de Comunidades de Agrarias de Guerrero (1933). The rich landowners in the state fought bitterly against the implementation of land reform, especially during the key years of the Cárdenas presidency (1934–1940), which witnessed the most dramatic levels of redistribution. For instance, 542,529 hectares of land were redistributed during 1934–1940, compared to 332,642 in 1920–1934, and 190,165 during the period 1940–1947.[27] In anticipation of armed attacks from *caciques*, Guardias Rojas were created during the Cárdenas administration to protect *campesinos*. These peasant self-defense groups roughly foreshadowed the creation of Policías Comunitarias (PCs) in the 1990s, as we shall see. As a premonition of horrific struggles that would appear throughout the remainder of the century, 26 *campesino* leaders were murdered in Guerrero in 1938 at the hands of *pistoleros* hired by *caciques* enraged by the land reform process.[28] Generally, though, Guerrero was most stable politically precisely during 1930–1960, when the process of land reform was in full swing, when the federal government enjoyed years of economic boom that financed the provision of agricultural credits and other social programs, and when the power of *caciques* was offset to some extent by land reform and government subsidies.

This period of stability quickly unraveled beginning in the 1960s, and introduced a political culture characterized by escalating repression and violence. During this decade, social protests were aimed at a failing agricultural sector as well as at the *cacique* system. These public demonstrations concentrated not only on the elements of political exclusion and authoritarianism, but also upon the system's punishing material implications that were noted earlier. Further, *caciques* dominated the lucrative contraband market, yielding a plethora of social and strategic implications. In the *cacique*-controlled illicit sector, which in the 1960s was more or less limited to the illegal harvesting of wood and not yet dominated by the narcotrafficking industry, peasants witnessed the resultant ecocide. By the early 1970s, about 80 percent of forestry in Guerrero was in the hands of four large companies, which often engaged in illegal logging.[29]

Peaceful social protests against this arrangement were often met with crushing repression. Notable examples include the 1960 murder by the military in the state's capital of Chilpancingo of 15 *campesinos* calling for the resignation of the governor at the time, Caballero Aburto, who they claimed acted as a *cacique.* The formation of the Asociación Cívica Guerrerense in 1962, a group that opposed this exclusionary system, was greeted with the murder of seven of its members at the hands of government forces. Adding to this volatile scenario, socialist organizations inspired by the Cuban Revolution proliferated during this period throughout Latin America, and Guerrero was no exception.

Lucio Cabañas, Guerrero's celebrated rebel hero who succeeded in reigniting the spirit of Zapata, made a strong political impact in the state beginning in 1959. Beyond his talents of political organization that centered on teachers and students, much of his work in the 1960s focused on a critique of an emerging agricultural crisis in Guerrero. The context was low prices for crops such as coffee, and insufficient credit and state support for *campesinos,* in tandem with the ill effects produced by illegal logging largely by foreign-owned lumber companies. Cabañas was detained by authorities on a number of occasions, as he was in 1960, just after the organization of the radical Frente Revolucionario Zapata, in which he played a leading role. He consistently led public protests, such as the 1962 blockade of a logging road leading to Sierra de Atoyac. Cabañas' work paralleled that of a notable compatriot, Genero Vásquez, who directed the Asociación Civica Guerrerense beginning in 1962. He, too, was arrested by authorities several times, and in 1963 publicly proclaimed that the *campesino* struggle in Guerrero must take a revolutionary path because there was no space for them within the existing Mexican political system.[30]

Revolutionary activity turned a corner in 1967 when Cabañas led a group called the Partido de los Pobres (PDLP)—the Party of the Poor. Allies included the Partido Revolucionario Obrero Clandestino Union del Pueblo (PROCUP), an organization that fused with the PDLP in the early 1980s and became the nucleus of what would emerge in 1996 as the EPR guerrillas.[31] Other comrades included members of the Liga Leninista Espartaco, the Partido Comunista Mexicano, and la Central Nacional de Estudiantes Democráticos.[32] Cabañas transformed from his position as leader of peaceful protests to a newfound role as guerrilla commando just after he led a rally in Atoyac de Álvarez to oust the conservative principal of the school at which he taught. Hundreds attended the demonstration, which served as a lightning rod for a popular protest against the *cacique* system and a budding agricultural crisis. Tragically, this rally on 18 May 1967 was met with a massacre on the part of military forces, which set the ball rolling toward the Dirty War. Cabañas, along with his allies, fled to the nearby mountains and proceeded to engage in guerrilla warfare until his death during combat in late December 1974.

Supporters of his guerrilla movement included *campesinos* in Atoyac de Álvarez, Coyuca de Benítez, and Tecpan, as well as indigenous members located principally in La Montaña.[33]—all key recruiting grounds for the EPR guerrillas in the 1990s and beyond. While Cabañas attracted aboriginal supporters, indigenous politics was not central to his PDLP. This was a peasant insurgent force. What did Cabañas and his supporters stand for? Cabañas observed in 1968, the same year of the student massacre in Mexico City, that "[w]e unite against the wood company, against the government, against starvation, against big business."[34] The problems that he identified—the illegal harvesting of wood, an exclusionary government, grinding poverty, and social inequity, as well as the need for further land redistribution and agricultural credits—are among those that remain at the core of social struggle in Guerrero in the new millennium and that have motivated a new generation of guerrillas. He repeatedly criticized the *cacique* system as key to the state's social malaise.[35] Cabañas attempted to shine the floodlight of justice on the atmosphere of violent repression that included the assassination of union leaders and other activists who demanded social change.[36] He was not afraid to direct his critical observations toward compatriots. He mocked, for example, the impracticality of leftist academics who devoted themselves to theory and then tried to "impose" it without fully "understanding the people."[37]

Cabañas and his party were influenced by a number of revolutionary thinkers.[38] In his various speeches, he resurrected the spirits of Hidalgo and especially Zapata. He also cited the heroism and strategies associated with Nicaragua's Sandino, along with those of Mao Tse Tung, Salvador Allende, and others.[39] With frequent reference to his compatriot Che Guevara,[40] Cabañas preached what he viewed as the unfortunate necessity of violent struggle. He lamented that his followers "were tired of peaceful struggle that obtained nothing."[41] Indeed, the urgency of Cabañas' tone was inspired by Che's famous assertion that "[i]t is not necessary to wait until all conditions for making revolution exist; the insurrection can create them."[42] Cabañas and his party also embraced Che's notion of the prolonged war.[43]

There are a number of noteworthy points to be made regarding Cabañas' campaign. First, his role as a radical schoolteacher is significant in that it relates to his views regarding education and political organization. That is, there is a kinship here to Freire's classic *Pedagogy of the Oppressed,* which advocates a "pedagogy (that) makes oppression and its causes objects of reflection by the oppressed, and from that reflection will come their necessary engagement in the struggle for their liberation."[44] Cabañas viewed education as an instrument of radicalization and ideological rebirth. He decried mainstream education of the time, which he portrayed as a vehicle for the ideas of the rich and for the repression of the majority population. Although Cabañas' attempts to radicalize students were limited to his town of Atoyac, the general principle of schools as sites for radical political organization would reach full bloom in Latin America during the 1980s and early 1990s with the Peruvian guerrilla group Sendero Luminoso.[45] Beyond his views on education, Cabañas seemed to hint at what later became Liberation Theology. He suggested to his devout Catholic followers that "there are saints within Christianity that are not enemies of the Revolution," and that Mexican revolutionaries had their own political saints in the form of Zapata, Villa, Hidalgo, and Morelos.[46] He underscored that "[w]e believe in God and our program is not absolute or dogmatic."[47] And with regard to his grand strategy, although Cabañas viewed Guerrero as the nucleus of his struggle, he clearly hoped his efforts would spawn a national revolutionary movement.[48]

There are, perhaps, some unflattering references to be drawn from Cabañas and Vásquez, as well. These have to do with their Robin-Hood-like justification of crimes such as kidnapping and robberies to support their revolutionary endeavors.[49] In August 1967, for example, Vásquez' group robbed a wealthy coffee plantation, kidnapping its owner and killing one of his workers. In 1971, the same group kidnapped the rector of the Universidad Autónoma de Guerrero.[50] In language reminiscent of the Colombian guerrillas' terminology of "taxes" for what others deem to be extortion, and "retentions" for what is widely viewed as kidnappings, members of Cabañas' PDLP referred to the 1972 "expropriation" of $230,000 (US) during one of their successful bank robberies.[51] Cabañas also hinted at what may be a dual message when he suggested that the government referred to his group as "poppy growers from Sierra de Atoyac."[52] Was he implying some relation between the group and narcotrafficking, similar to the allegations made against the EPR in the 1990s and into the new millennium? Was he suggesting, instead, that such dismissive brandings were simply a way for authorities to shrug off the revolutionaries as common criminals?

Phase Two: Denial and the Dirty War—State Terror and Genocide

> There's nothing to worry about. . . . Those people don't really count.
>
> *Juan Rulfo,* Pedro Páramo

While participating in what the Mexican government in 2006 deemed to be state terror and even "genocide" as defined by the United Nations,[53] it is remarkable that until 1971 Mexico's Secretaria de la Defensa "systemically negated" the existence of any guerrilla group in Guerrero.[54] Despite this claim, the government by that year had assigned 24,000 soldiers, or about one-third of the country's entire army, to the state of Guerrero to eliminate guerrilla forces. Intense military operations associated with the Dirty War had actually commenced a year earlier, with the launching of Operación Amistad between 25 July and 13 August 1970. This exercise entailed arbitrary detention, torture, sexual assault on women with their husbands present, and possible extrajudicial executions, among other tactics, according to the government's 2006 report on the Dirty War.[55] With the death of guerrilla leader Genero Vásquez in a car accident in February 1972, and the almost immediate dissolution of his group of followers, the military focused its efforts on the larger and more important movement of Cabañas and his supporters. By 1972, massive detentions on the part of the military were routine,[56] a practice that was highly abusive to the majority of those detained who happened to have no association with the guerrillas. The clear purpose of this practice was to terrorize the population to prevent them from lending support to the revolutionaries.

The government's official report on the Dirty War documents a triad of strategic tactics on the part of the Mexican military that amounted to "brutal repression of the population." These included the army's attempt to instill intense fear into populations suspected of supporting guerrillas, the mere pretension of punishing army members guilty of abuse, and the routine application of torture on detainees.[57] Military bases such as the one in Pie de la Cuesta, a tourist beach just 20 minutes north of Acapulco, transformed into "centers of clandestine detention" for abusive treatment of detainees, according to the report that lists the full names of hundreds of victims.[58] Fishers working in the ocean off Acapulco in 1972 and 1973 regularly discovered the clothes and other personal articles of those victims who apparently were pushed out of military helicopters. The fishers subsequently were threatened by the army not to make public disclosures regarding their gruesome discoveries.[59] Given the extraordinary military pressure exerted in rural areas of Guerrero during the Dirty War, Cabañas during this period considered taking the revolutionary struggle from the countryside to Mexico's big cities. He contemplated sending members of his group to Spain to learn techniques in urban warfare from that country's insurgent group the Euskadi ta Askatasuna (ETA).[60] With his days numbered, Cabañas did not have time to put that idea in motion, and Mexico was not subject to the urban warfare faced by Peruvians in Lima or Colombians in Bogotá, Medellín, and Cali.

Cabañas' PDLP began planning in January 1974 the kidnapping of Senator Rubén Figueroa, who proceeded to become the state's notoriously repressive governor. He was kidnapped by Cabañas' group in the late spring of 1974 and rescued 100 days later, on 8 September. During that period, 63 military specialists from Brazil's Escuela Superior de Guerra (Superior War School) arrived in Guerrero to assist the Mexicans with Brazil's own expertise in directing a grim Dirty War.[61] About three months after Figueroa was liberated, amidst redoubled efforts by the military, Cabañas died in El Otatal on 2 December 1974 during a confrontation with the army. Cabañas' Party of the Poor virtually disintegrated soon after his death, unable to sustain itself without a charismatic and visionary leader.

But this did not halt the Dirty War. In fact, in an effort to rout any remnants of the guerrillas, military repression continued in full tilt after Figueroa became governor in April 1975, and during the presidencies of Echeverria and López Portillo. This period overlapped with the commencement of daily intelligence reports by the CIA.[62] Massive detentions and a "war of extermination" were the norms in Guerrero until the early 1980s, according to the government's official report, which meticulously documents the death of at least 650 persons who "disappeared."[63] Governor Figueroa in 1980 commented that "[h]ere there are no disappeared, which we call communists. . . . Disappeared? There're all dead!"[64]

Cabañas' guerrillas were not alone in Mexico during the 1960s and 1970s. Indeed, Mexico's intelligence agency estimated the presence of 29 rather small guerrilla groups during this epoch.[65] But it is the ghost of Cabañas himself that continues to haunt Guerrero and other areas of southern Mexico such as Oaxaca and Chiapas. His name has been frequently invoked by social movements, and by guerrilla groups such as the Movimiento de Acción Revolucionaria in the 1970s and the EPR and its offshoots in the 1990s. Chiapas' EZLN have also made reference to the struggle he initiated.

The Dirty War occupies a crucial and highly disturbing place in Mexico's strategic history. Some three or four decades after that, the Fox government, to its credit, commissioned an extraordinary report that documented in painstaking detail what amounts to state terror, especially in Guerrero. Those were the days of the closed nation-state. A government could get away with clandestine warfare against its own population, and even commit a massacre of students in Mexico City on the eve of hosting the Olympics, and not have to endure the heat of serious public condemnation. What is also striking is that so many of the problems identified by social forces during that era persist these days in a manner that is linked to a plethora of strategic issues apparent in the new century. Guerrero has been underexamined in relation to security themes, and the current social struggle must be placed within the historical context of the Dirty War and the trajectory that led to it.

* * *

Beyond the extraordinary militarization of Guerrero in efforts to eliminate leftist guerrillas, the government in the 1970s also responded with a significant aid package for *campesinos* in radicalized areas of the state. This included 200 roads, 90 hydroelectric projects, and the augmentation of agricultural credit systems especially for the essential crops of corn and coffee.[66] This classic carrot-and-stick tactic would be mimicked by the Mexican government in future struggles, such as in Chiapas in the late 1990s. While the government wished to be applauded for its much-needed assistance to peasants during the Dirty War, it also sent a disturbing subtextual message: violence begets social programs that would be absent otherwise. Despite the provision of such programs, small numbers of committed revolutionaries who would later form the nucleus of the EPR guerrillas continued to organize. These included members of PROCUP and sympathizers of Cabañas who survived the military's assault, and who forged ahead in their efforts to organize marginalized workers, peasants, and students. Other groups fell into the fold, such as the Frente Estudiantil Revolucionario, the Federación de Estudiante de Guadalajara, and the Liga Comunista 23 de Septiembre.[67]

While some *campesinos* benefited during the 1970s from the provision of government programs, another force was well underway that economically detracted from Guerrero's rural residents. Acapulco, which blossomed into an important tourist destination beginning in 1946, when it won the favor of the Hollywood elite and even Howard Hughes, was increasingly being viewed by the government of Guerrero as the state's most significant cash crop. The quest to increase tourism in Acapulco drew more than half of the funds allocated to municipalities in Guerrero beginning in the 1960s.[68] A growing resentment emerged among impoverished *campesinos* regarding what they viewed as lavish government funding for a playground of the rich.

Phase Three: Guerrero and the Imposition of Neoliberalism

The dual-pronged government strategy of intense repression, mixed with new social programs that predominated during the 1970s, gave way to a scenario of economic calamity following the 1982 debt crisis. At the national level, both poverty and inequity worsened after 1982 and into the 1990s. In 1984, the income of the poorest 10 percent of Mexican families constituted 1.72 percent of total Mexican GDP, and by 1989 it fell to 1.58 percent before cascading to 1.55 percent in 1992. By 1996, the bottom 50 percent of the population received only 10 percent of the national wealth. At the other end of the spectrum, the top 10 percent of the population watched their share of wealth increase from 32.77 percent of total Mexican GDP in 1984 to 37.93 percent in 1989 and to 41 percent in 1996. IMF restructuring plans spelled less support for the poor during an epoch of crisis. Government spending per capita fell 3.5 percent in 1982 and a huge 31.5 percent in 1983. The share of health-care spending in the GDP fell from a predebt figure of between 3.5 and 3.7 percent in 1980–1982 to an average of 2.85 percent between 1983 and 1988. A similar trend can be found with respect to spending on education as a share of GDP, which fell from 3.8 percent in 1982 to about 2.6 percent between 1985 and 1988.[69]

Of particular concern in southern Mexico was a protracted clawback of longstanding government support vital to agricultural production. While Guerrero certainly benefited from land reform more than, say, Chiapas—between the years 1900 and 1992, a total of 4,998,630 hectares of land were redistributed in Guerrero in comparison to only 1,076,276 in Chiapas—such reforms depended strongly on government support programs such as agricultural credits, marketing and purchasing agencies, and so on. But by 1983 in Guerrero, one year after the emergence of the debt crisis, government spending on agriculture receded to less than the level in 1977, and continued to slump in subsequent years. For example, a key government agency, Conasupo—which assisted in the planning, regulation, and marketing of crucial crops such as corn, beans, and rice—saw its budget cut by 61.2 percent between 1983 and 1987. Similarly, Banrural, a government bureau that provided credit to peasants, offered its assistance to about 6,000 Guerrero coffee growers in 1979, which fell 75 percent to just 1,500 in 1983.[70]

The point is that the economic and social effects of the debt crisis meant that conditions for Guerrero's *campesinos* were even worse in the 1980s than during the era that prompted the formation of Cabañas' guerrilla group. Following the debt crisis, then, the economic context appeared ripe for the birth of new revolutionary organizations. There was some movement in this direction, with the temporary fusion in 1981 of PROCUP and what was left of Cabañas' PDLP.[71] In an account of its own history, the EPR guerrillas indicated in 2005 that during the 1980s authorities either "satanized" PROCUP or denied its existence.[72] At any rate, the fusion of the two small rebel bands had more to do with the sheer survival of these groups in the aftermath of the Dirty War than with any attempt to project power. Attempting to put the best face on a rather bleak predicament, comandantes of the PDLP in 1987 insisted that their coordination with PROCUP meant that the movement was "strengthening" and moving forward.[73]

The effects of the debt crisis ripened the political conditions that spurred the slow emergence of a new set of guerrilla groups during the mid-1990s. Resentment against the *cacique* system lingered. With the apparent fraud in the 1988 federal elections in favor of the PRI's Salinas and at the expense of the Partido de la Revolución Democrática (PRD)'s Cárdenas, Guerrero's *campesino* population grew increasingly restless. This was largely due to the PRI's domination not only of national politics but also of the Guerrero government, and its central role in the *cacique* system. Between 1988 and 1990, there were 56 assassinations of PRD supporters in Guerrero, 25 illegal detentions, and a wide assortment of other brutalities. This atmosphere of extreme exclusion and repression was exacerbated under the reign of the newly elected governor of Guerrero, Rubén Figueroa Alcocer—a *cacique* who was the son of the infamous Rubén Figueroa Figueroa, Cabañas' kidnapping victim and the notorious governor of the state during the height of the Dirty War, from 1975 to 1981.

According to its own literature, PROCUP-PDLP, the group that would later form the EPR, was ebbing in the early 1990s, when it faced "possible decomposition" and a "moment of ideological weakness."[74] Despite this, the EPR claimed that it was responsible for a key political development in Chiapas in 1992—the smashing in San Cristóbal of the statue of Diego de Mazariegos, viewed scathingly by the indigenous as a symbol of imperialism.[75] This is significant not only because the event was among the catalysts that sparked the 1994 EZLN uprising, but also because, if it is true, it suggests the projection of PROCUP-PDLP's power throughout key areas of southern Mexico. Further, the U.S. Department of Defense in April of 1992 observed that "[t]he Mexican Defense Secretariat is concerned about the insurgent group PROCUP-PDLP." This was due to the group's opposition to Nafta, its attempts to attract international attention, and its campaign for the sympathies of university students in Guerrero, Oaxaca, Michoacán, Morelia, and Puebla.[76]

During this period, economic conditions in the countryside worsened, not only in the wake of the 1980s debt crisis, but also as a result of Salinas' 1992 termination of land reform that had been enshrined in Article 27 of the Mexican constitution. Guerrero was hurt by shriveling government agricultural support programs, which resulted, among other things, in the sale of fertilizer falling 57 percent between 1987 and 1992. The third poorest state in Mexico, Guerrero, witnessed escalating levels of poverty. In 1990, for example, 50 percent of the state's residences had dirt floors, which rose to 57.6 percent in 1996; availability of running water fell from 57 to 50 percent during the same period.[77] Against that backdrop, rumors mounted regarding the formation of guerrilla groups, such as the Ejército de Liberación de la Sierra Sur, Movimientos Popular Revolucionario, Ejército Insurgente de Chilpancingo, Fuerzas Armadas Clandestinas, and the Ejército de Liberación del Sur.[78]

A decisive turning point occurred with the murderous ambush by Governor Figueroa's henchmen of 17 *campesinos* associated with the Organización de Campesinos de Sierra del Sur, a peaceful group that has advocated the rights of peasants. As we shall see in Chapter 6, this event was the catalyst that transformed PROCUP-PDLP into the EPR guerrillas. Let us proceed to trace historical elements of Guerrero's social struggle to recent manifestations of the RMA in the realm of political economy.

6

Fight or Flight

Insurgency and Migration, Surveillance and Complexity

In the last chapter we observed that a number of small resistance groups sprouted in Guerrero during the early 1960s to protest against an emerging agricultural crisis as well as the exclusionary features associated with the semifeudal *cacique* system. These groups were greeted by government repression that aimed at shrinking political space for the marginalized population, a move that contributed to the founding of guerrilla movements led by Genero Vásquez and the revered Lucio Cabañas. This sparked even more repression through brutal government efforts to exterminate the insurgency during the now infamous Dirty War, a disturbing era marked by flashes of state terror and genocide.

What is striking is that a rendition of that sad cycle occurred in the 1990s and continued into the new millennium. This era has been marked by the emergence of an agricultural crisis that has affected much of the country, and by growing signs of ungovernability especially where Mexican hegemony has worn thin on the periphery of the country—such as in the three poorest states of Guerrero, Oaxaca, and Chiapas. Unlike the national and highly violent insurgencies that have swept virtually the entirety of Colombia, Mexico's guerrilla movements have been localized in southern parts of the country. Beyond the effects of limited regional insurgencies, by 2007 prospects for broader national instability appeared to be magnified with the shaky legitimacy of the new Calderón government, an alarming wave of escalating violence associated with narco-warfare, a likely deceleration of the Mexican economy, and with the construction and fortification of walls along the border in the context of Fortress America.

This chapter will begin with a brief discussion of the context of neoliberalism that has framed the genesis of new insurgents and social resistance movements in Mexico. The emphasis will be on the classic dynamic of "fight or flight." That is, in the context of economic depravation and social injustice, what specific conditions prompt social forces to form resistance movements and fight the powers that exist, or catalyze social forces to take flight en masse and illegally migrate to the United States? To what extent can social forces work constructively and creatively with elements of the RMA to advance their interests, and to what extent are social forces victimized by the manipulation of the RMA on the part of the state?

An examination of insurgent warfare will be featured here. Using the same pattern of elements associated with the RMA that were explored in the Colombian case, we shall begin with a focus on the political instrument of fear that at times has morphed into terror. We shall then move on to a discussion of asymmetric dimensions of conflict. Next, the theme of complexity will be examined through a focus on the strategic implications of migration. There will be an analysis of the contradictory forces of economic integration through Nafta versus U.S. fears of terrorism. We shall also consider the proposed construction of a fence that conjures images of a new "iron curtain" that reverses the dynamic of its cold war counterpart—that is, this fortress is designed to lock people out rather than to prevent them from escaping. Related to this redefinition of the strategic meaning of the frontier is another key element of the RMA, the use of ultra-surveillance. Let us proceed to probe the social and economic context that has framed the emergence of Mexico's most recent RMA and that has prompted the dynamic of fight or flight.

The Neoliberal Ledger

> Now I perceive the absurdity and delusion of them, and am only sorry I am undeceived so late.[1]
>
> *Miguel Cervantes,* Don Quixote

There have been mixed economic results since Mexico became a Nafta partner.[2] The good news is that Mexico, since 1995 to the time of this writing, has not experienced a major currency meltdown or debt crisis.[3] This, indeed, is quite an accomplishment, especially given the severity of the debt crisis of 1982 and the "tesobono" disaster of 1994–1995. Mexico's majority population suffered enormously during these ordeals, in terms of spiked poverty, diminished employment, reduced social services, curtailed purchasing power, and so on. Further, the United Nations Economic Commission for Latin America and the Caribbean (ECLAC) notes that overall poverty rates in Mexico fell from 44.2 percent in 1992 to 37 percent in 2004. It is important to underscore that macroeconomic stability and minimal inflation rates achieved in Mexico represent major accomplishments that benefit both the rich and the poor. Adding to the positive side of the ledger, Mexico took the title of Latin America's largest recipient of foreign direct investment in 2005 at $17.8 billion, signaling strong global investor confidence in the country. And, by 2005, Mexico was the tenth largest exporter in the world.

Despite such kudos, there have been some glaring debilities in the Mexican economy. These have had a negative impact on security, especially in relation to illegal migration and the acceleration of narcotrafficking. There also have been mounting and raucous social struggles against government policies designed to cope with a shifting economic base. Let us proceed to examine some of the vulnerabilities in the Mexican economy, beginning with debt.

In many ways, Mexico has managed its debt well since 1995. External debt as a percentage of GDP fell from 33.7 percent in 1999 to 20.5 percent in 2004. But some other figures are less sanguine. Despite cascading levels of external debt, the cost of debt servicing nevertheless has been steep. Between January 2001 and May 2006, Mexico's interest payments on external debt totaled $61.7 billion, or about 37 percent of its total external debt of $165.4 billion in 2006.[4] Total public debt represented 40 percent of the country's GDP in that year, a level considered "high" by the chief economist at Mexico's finance secretariat.[5] And while the country has improved its insulation from debt shock to some extent, with about two-thirds of its debt owed in pesos rather than in dollars by 2005 compared to only one-third in 1998,[6] a sharp spike in interest rates easily could precipitate another crisis. This is true not only at the national level, but even at the conglomerate personal level, given that personal debt rose 46 percent in 2005 alone.[7] In the past, debt crises have been accompanied by elevated levels of illegal migration and the augmentation of the illicit economy.[8]

Further, Mexico's GDP per capita has remained nearly flat between 1980 and 2004, compared, for example, to over 350 percent growth in South Korea's economy.[9] Thus, Mexican living standards have stagnated. Not only has growth been lackluster in Mexico since the 1980s, but social inequity also worsened during that same period. Mexican gini coefficients rose from 0.47 in 1984 to about 0.50 by 2005.[10] Crucially, ECLAC estimates that much of the reduction in poverty to which we referred above is due to sharp increases in the levels of remittances sent by illicit migrants to families in Mexico. ECLAC statistics suggest that in 2002, for example, 45 percent of Mexicans lived in poverty, but that poverty rates would rise to about 65 percent if remittances were not included.[11] Insufficient growth, poverty, and rising inequity have combined to provoke social tensions and related security issues, as we shall see.

Not only does anemic growth mean stagnant living standards, but it also spells a serious incapacity of the economy to absorb hoards of would-be employees. The Mexican economy grew 3 percent in 2005 and 4.8 percent in 2006, with a projected deceleration of the economy through at least 2008 due to a slowdown in the United States. To accommodate the nearly 1 million new job seekers who appear annually, the country would need to revisit the 6 percent yearly growth rates that Mexico enjoyed during its heyday of the 1930s to the 1960s. President Fox, early in his tenure, promised 7 percent growth rates and the creation of 1.4 million jobs on an annual basis.[12] But these expectations have been missed by a long shot. Not only has the country failed to witness the economic acceleration that could create desperately needed employment, but there has actually been a net loss of 2.1 million jobs in Mexico between 2000 and 2005. About one-third of Mexico's working population has been forced into the so-called informal sector, while millions of others have resorted to the safety valve of illicit migration to the United States, a point to which we shall return.

Let us explore the Mexican employment predicament a bit further, with a brief consideration of three particularly important areas—manufacturing, agriculture, and tourism. Central to Mexican manufacturing is the maquiladora sector, which comprises export-oriented factories concentrated near the U.S. border. Americans owned over three-quarters of these factories in 2006. Although *maquilas* were meant to bolster Mexico's industrial output, the country's industrial sector actually shrank from 34.9 percent of total GDP in 1984 to 26.4 percent in 2004.[13] Further, the *maquilas* did not yield the employment opportunities initially envisioned. The number of *maquila* jobs peaked in the year 2000 with about 1.3 million employees, falling to 1.17 million by late 2005.[14] Much of this has had to do with Mexico's growing incapacity to compete with China and other countries in light manufactured products.[15] Reflecting that trend, average wages in the *maquilas* tumbled 57 percent during the same period.[16] The IMF suggested that in order for Mexico to cope with increased Chinese competition in the manufacturing sector, it must react by "easing labor market rigidities," or, in other words, by cutting wages.[17] But shrinking paychecks are likely to fuel illicit migration. Significantly, the total number of *maquila* employees in 2005 roughly equals the number of Mexicans who migrate to the United States in just a two- to-three-year period.

Compounding the employment crisis in the Mexican manufacturing sector have been mounting problems in the realm of agriculture. An estimated 900,000 to 1.3 million agricultural jobs have been lost, largely in southern Mexico, between Nafta's commencement in 1994 and the year 2004.[18] As a percentage of total GDP, this sector fell from 9.4 percent in 1984 to 4.1 percent in 2004. The political economy of corn is a big part of the problem. Although this crop is just one factor, and while there are indeed corporate farms in the country that have flourished by cultivating certain vegetables and niche products, the place of corn in the Mexican diet and culture renders it a lightning rod for broader problems in the agricultural sector. Massive quantities of highly subsidized corn from the United States, the world's largest cultivator of this product, have been dumped in Mexico at prices even lower than those charged by peasants struggling to subsist. The United States subsidized corn production costs at a rate of 37 percent between 2000 and 2002.[19]

U.S. exports of corn to Mexico skyrocketed against the backdrop of high U.S. subsidization as well as Nafta provisions that have reduced agricultural tariffs. They jumped from 0.5 million tons in 1993, the year prior to Nafta's implementation, to 10.7 million tons by 2006. More than 90 percent of the farmers in the state of Guerrero have cultivated corn and have been deeply affected by the displacement of their produce by American corn.[20] A new wrinkle in the political economy of corn occurred in early 2007. Corn prices spiked as a result of rising ethanol production in the United States, which sparked a political crisis in the form of sharply higher prices for tortillas, the staple of the Mexican diet. President Calderón intervened in January 2007 to cap tortilla prices in the context of widespread public protest by the poor, and also lowered quotas to permit the importation of even greater quantities of U.S. corn. While what was left of Mexico's corn farmers would have benefited from higher prices, the tortilla crisis was rooted in part in Mexico's increasing and overwhelming dependence on imported U.S. corn since the implementation of Nafta. At the time of this writing, no government programs were in place to address the issue of corn cultivation or the wider agricultural crisis. Overall, the failure of Mexico's agricultural sector since the implementation of Nafta, especially in areas such as corn cultivation, has propelled illegal migration to the United States as well as the cultivation of illicit drugs such as opium and marijuana. It has also provided fodder for armed groups in Mexico that have been critical of the country's direction since 1994—points that will be developed in the subsequent chapter.

To a limited extent, Mexico has been able to compensate for the loss of jobs and revenue in the manufacturing and agricultural sectors by growth in the tourism industry. After oil and remittances, tourism has been Mexico's third largest source of foreign revenue—estimated at about $11 billion in 2005.[21] Mexico has been Latin America's leading tourist magnet, with about 70 percent of the country's visitors coming from the United States. The state of Guerrero attracted the largest portion of tourist investment in the country in 2004 ($429.9 million), and so that state represents an important microcosm from which to view components of this industry. Acapulco, less than a five-hour freeway drive from Mexico City's 22 million inhabitants, along with Zihuatanao and Taxco, are Guerrero's principal attractions. The question for our purposes is the extent to which the tourist sector in Guerrero, and especially in Acapulco, is related to security issues entailed in the RMA.

Some of the biggest and most profitable hotels and tourist attractions in Acapulco are foreign owned, such as Fairmont's Acapulco Princess and the Hyatt Regency. Given that tourism represents about 70 percent of the state's economy, it is noteworthy that the business class in Acapulco, and therefore in Guerrero, has been politically weak partially due to foreign dominance of the sector. This weakness implies that the state of Guerrero has enjoyed relatively little political leverage with the national government, which limits its ability to attract federally funded projects that could ease escalating social discontent.[22]

Viewed within the wider national context, in 2005 there were about 1.8 million employees in the country's tourism industry—or about the same number of illicit migrants entering the United States over a four- to five-year period. Given the backdrop of a vast army of surplus labor, wages in this sector have been generally low—about $8 to $10 for a day's work in 2006 for nonmanagerial and nontipped positions. Compensation, then, has not been sufficient to diminish the allure of illegal migration to the United States or to deter the monetary attraction of the illicit economy, such as narcotrafficking.

Further, there have been instances in the late 1990s and the early part of the new millennium when Acapulco's tourism industry was threatened with ecocide. One such inflection occurred in 2003. With no sewage treatment plant, and thus raw sewage pouring into the ocean from the city's more than 1 million residents, in addition to riverine sewage and other wastes streaming into the sea, ocean contamination has been considered to be "extreme" by academic scientists.[23] One expert told a U.S. news agency that "[w]hen we brought up the problem in 1990, we were called communists and anti-capitalists."[24] Much of this became public knowledge in 2003, when the Mexican environmental secretariat banned the use of two beaches in Acapulco during Easter holidays due to obvious threats to human health. But, extraordinarily, the decision was vetoed by Acapulco's mayor with the support of the local archbishop.[25] The highly publicized scandal prompted a temporary decline of up to 25 percent in tourism. Beyond the concerns regarding sewage-related fecal coliforms, water has been polluted by other toxic waste and by the obvious presence of everyday garbage.[26]

Overall, the tourism industry in Guerrero is fraught with challenges related to security concerns associated with the RMA. The tourism sector has not been able to compensate for the weaknesses in the agricultural and manufacturing sectors. Workers in these playgrounds for the rich are generally not paid well, leaving many to ponder the option of material incentives associated with illicit migration. Further, the tourism sector in Acapulco has been under siege with threats emanating from ecocide and also from violence linked to the city's burgeoning narcotrafficking industry. The latter caused serious concern in early 2006, when a spate of narco-related terror may have contributed to a steep decline in hotel occupancies during Easter week in Acapulco.[27] Escalating violence in that year and in 2007 led to expressed concern at the highest levels of Mexican politics regarding its negative impact on the vital tourism industry, and prompted President Calderón to note that "more security will mean more tourism."[28] The tourist sector in Mexico, then, is related to the RMA with respect to themes such as global crime, ecocide, and migration.

* * *

It is worth emphasizing that Mexico's economic woes have been glaringly obvious and have been recognized by the country's elite. They have understood that trouble in Mexico's agricultural and manufacturing sectors has not been offset by the service sector, which constituted 69.5 percent of the country's total GDP in 2004. Carlos Slim, Mexico's richest person and one of the wealthiest people in the world, put forward a plan in February 2006 that aimed at achieving vibrant growth rates and reduced inequity. His Plan for National Extension and Citizen Inclusion focused on exploiting Mexico's raw materials and abundant labor. Declaring that 3 percent growth in 2005 is insufficient, he aspired to augment that rate to the 6 percent standard enjoyed in the middle of the twentieth century. But that achievement had occurred in a different world, when a nationalistic Mexico still had a strong state and did not have to compete with race-to-the-bottom wages that are sometimes paired with globalization. Further, in that bygone era Mexico did not have to cope with some harsh realities associated with Nafta, and with the discipline imposed by finance capital that can drain billions from the economy in a matter of nanoseconds.

Slim's plan was important to the extent that it signaled a clear recognition on the part of Mexican capital that the country's mix of inequity and stalled growth represent a harbinger of crisis. In fact, in 2004, Slim noted that because the neoliberal model has been applied for some years and has not resulted in increased employment, there exists a risk that Mexico will embrace populism—a clear reference to the demonized Hugo Chávez. Hence, Slim's interests and those of his followers were not simply philanthropic, but reflect the well-founded preoccupation that Latin America's tilt toward the Left will reach Mexico and rescind the privileges of the country's proportionately tiny elite.

Overall, a central problem in Mexico since 1982 has been rising inequity combined with a severe lack of employment opportunities for those desperate to work. In notable cases this predicament has been exacerbated by government repression. To a significant extent, then, Mexico's security problems are a reflection of a flawed model of development. This has framed the phenomenon of fight or flight. Some social forces have chosen to "take flight" and attempt the increasingly perilous and expensive trek to the United States in an effort to find work. Others have chosen to remain in Mexico and fight for social justice, and this is where our analysis shall begin, turning first to those who have chosen armed struggle rather than peaceful endeavors.

Mexico's Military Insurgencies: Asymmetric Warfare and the Discourse on Terror

> Make him pay, son, for all those years he put us out of his mind.[29]
>
> *Juan Rulfo,* Pedro Páramo

The reliance on force by Guerrero insurgents is a central feature that has differentiated them from the EZLN in Chiapas. While the Zapatistas have at times wielded a national influence in Mexico due to the power of their ideas and their novel means of dissemination, Guerrero's guerrillas have been isolated as a result of their penchant for violence and for a generally outdated ideology and strategy. This dichotomy has been exploited by the government in its distinction between "good" guerrillas such as the nonviolent Zapatistas versus the "bad" or even "terrorist" guerrillas present in Guerrero and elsewhere in southern Mexico.

While Chiapas and the EZLN rightfully have received due academic attention, Guerrero's guerrillas have been largely ignored. Part of their relative obscurity has to do with the government's demonization of them through the discourse on terror, and much of it has had to do with the EPR's ineptitude to navigate successfully through dimensions of the RMA. Other peculiarities of Guerrero, compared to Chiapas, include the continuity of the struggle since the 1960s, the clandestine and closed nature of the groups involved, and the lack of a singular and massive uprising to herald the arrival of a truly revolutionary force. Guerrero's insurgents also have been characterized by the absence of a charismatic figure, by the fragmentation of those engaged in armed struggle, and especially by the failure to harness traditional or original ideals to a strategic mind-set commensurate with the times. The relation between the RMA and the insurgents will be considered in three phases, beginning with their origins, turning to a subsequent period of fragmentation, and finally to the era of post-9/11 withering.

Phase One: Insurgent Origins, 1996–1999

The question of "how to tell the story" of politics is fundamental, because the philosophical principles underpinning that process are also reflected in the objectives and strategies of the speaker. So, an interesting point of departure concerns the fabrication uttered by the EPR itself regarding the story of its origins. In a September 2005 document titled "A Little More History," the EPR revealed that in

1995 it had "invented" a story that it circulated, which claimed that the group was an amalgamation of 14 distinct organizations. This was due to "tactical reasons" that were designed to lead Mexican authorities astray. The rebels hoped government forces would disperse their efforts in pursuit of several phantom organizations, rather than concentrate on the actual structure of the EPR, which was a fusion of the PROCUP and PDLP.[30]

There are a number of implications that flow from this. First, the clandestine nature of the group has afforded it the latitude to invent tales that have gone unquestioned by researchers and authorities. Rather than the limelight occupied by Chiapas' ELZN, where those rebels have placed themselves under public inspection and scrutiny, the EPR has resided in dark shadows that obscure accurate knowledge regarding the group. Much of the limited information that does exist on them has been disseminated by the rebels themselves. Indeed, until 2005, key published articles and other discussions on the EPR began with the falsity that the group was originally composed of 14 separate entities. Second, given that the EPR has expressed frustration regarding its inability to attract much public attention inside and outside of Mexico, a point to which we shall return, revelations that the group disseminates lies are not likely to remedy that situation. Public support outside their constrained geographic region depends in part on truthful and accurate information that global civil society can embrace. Finally, this prompts one to guess what other falsehoods the EPR may have propagated. For example, beginning in 1998, there have been numerous reports that the group splintered into several other guerrilla organizations—were these, too, all a tactical lie? We shall consider the evidence for this.

Let us consider the circumstances under which the EPR emerged. In terms of organizational structure, the EPR was originally headed by six comandantes—Antonio, Jose Arturo, Francisco, Vicente, Oscar, and a woman, Victoria. Authorities indicated that their troops numbered under 500,[31] but local experts suggest their ranks were much larger than that, perhaps a couple thousand, especially during their most potent period of 1996–1998.[32] The rebels' inaugural document, the "Manifesto de Aguas Blancas," was released on 28 June 1996 to commemorate the massacre one year earlier of 17 members of the Organización de Campesinos de Sierra del Sur (OCSS) who were ambushed and murdered by government forces—a point to which we shall return in the subsequent chapter. OCSS leaders stated that the Aguas Blancas massacre was the decisive event that catalyzed the formation of the EPR, against the backdrop of severe economic and social difficulties noted earlier.[33] Referring to the massacre and the long legacy of repression in Mexico, the EPR observed that "[i]n the face of institutionalized violence, armed struggle is a legitimate and necessary resource for the people to regain their will and sovereignty, and to reestablish their rights" against the backdrop of "decades" of "reactionary violence of exploiters and oppressors."[34] Given that the Zapatistas were militarily toothless just 12 days after their emergence in January 1994, the EPR's reliance on armed struggle immediately differentiated them from the EZLN.

In its original manifesto, the EPR stated five principal goals. First, "we struggle to overthrow the anti-popular, anti-democratic, demagogic and illegitimate government." Second, they endeavored toward "the reinstatement of popular sovereignty and reinstatement of the fundamental rights of man," a comment meant to emphasize that their new government would tame what they viewed as the pernicious forces of national and transnational capital. Next, the EPR emphasized its commitment to the "struggle for resolution and fulfillment of the immediate needs" of the people. The rebels indicated that they wished to establish international contacts, in contrast to the reclusive nature of groups such as Peru's Sendero Luminoso: "We struggle for the establishment of just international relations with the international community." Finally, they underscored their commitment to fight against "political oppression, misery, hunger, and crimes against humanity."[35] It is worth emphasizing that the EPR employed classic Marxist analysis, pointing to the exploitive forces of capital vis-à-vis "simple men and women." Thus, the broad features of class forces rather than the more specific elements of identity politics associated with the RMA, such as race, gender, and religion, were emphasized by the EPR. This stood in sharp contrast to their postmodern counterparts in Chiapas.

A more refined document emerged in August 1996, and was titled the "Manifesto de la Sierra Madre Oriental." One passage in it refers to the "imposition of the bourgeoisie, the oligarchs, and the antipopular government for 70 years." This is significant because the EPR did not consider the major social problems facing *campesinos* to be recent in origin, but viewed them as related to a legacy of failure to achieve ideals established in the Mexican Revolution. Wanting to believe that most Mexicans felt as the rebels do, the group initially embraced a grandiose scheme to mobilize "the masses" through classic Marxist-Leninist struggle.[36]

In a wide-ranging but not always coherent presentation, the EPR made four principal demands in this manifesto: the creation of a new provisional government, the establishment of a new constitution, the construction of popular democracy, and an economic reordering involving a major redistribution of wealth.[37] Besides these main issues, the document presented a wish list of 45 smaller points. These ranged from the renegotiation of Mexican debt, the establishment of migrants' rights, better agricultural programs, and respect for indigenous autonomy, to demands for unlimited public access to sports and physical education, the promotion of science and technology, and support for the arts. In striking contrast to the Zapatistas' treatises, this document clearly was not authored by a man of letters such as Subcomandante Marcos, who at one point flatly told the EPR, "We do not want your support, we do not need it."[38] Although the manifesto cited many of the problems the EZLN had identified, it did so without the statistics, the artistic verve, the literary shine, and the engaging hooks that have been instrumental to the success of the EZLN. The EPR is a homegrown *campesino* group in Guerrero that has desperately wished to help the poor but it clearly has lacked the capacity to navigate the postmodern media and communication systems.

Also in August 1996, Comandantes Vicente and Oscar, in an interview with a Mexico City newspaper, made what the government considered to be an alarming announcement—that the group had placed "fresh forces" within Mexico City. The EPR attempted to present itself as a national phenomenon, and not confined to a specific region as the EZLN had been. The rebels also indicated that they were coping as best as they could with the mobilization of what they estimated to be 23,000 Mexican military troops working to contain them.[39] Many of these were dispatched to both coastal and mountainous regions of Guerrero, and also to certain enclaves elsewhere, especially in Oaxaca. Three days after the interview, on 28 August 1996, the rebels conducted coordinated strikes in six states, causing 14 deaths and injuries to 23 others. By this point, there was no question that the EPR had represented a truly national concern, though Guerrero, and to a lesser extent Oaxaca, were clearly its chief bases of operation. The government's response was intensified and often indiscriminate repression, which at times was expressed as state terror.

One of the most notorious manifestations of this amplified repression occurred in Loxicha, Oaxaca, where the military and police arrested the entirety of the town's municipal authorities in September 1996 and initiated a lengthy campaign of what residents and human rights agencies describe as sheer terror. The leader of the Organización Pueblos Indígenas Zapotecas, who was imprisoned for four months in a jail in Almoloya in addition to being falsely accused of belonging to the EPR, suggested that government repression was "designed to terrorize the population from organizing, and to fragment the population politically."[40] Leaders of other human rights NGOs in Oaxaca echoed the view that the heavy-handed government repression in Loxicha and other regions was meant to create "fear and division" among the population so as to prevent another uprising on the scale of the Zapatistas in Chiapas.[41]

Although government repression would eventually take its toll on the rebels, by the end of its inaugural year of 1996, the EPR presented itself as a vital and growing insurgent group. It claimed to be present in 17 states, but was clearly noticeable in Guerrero, Oaxaca, Veracruz, Chiapas, and Hidalgo. Further, in a show of armed strength, the EPR in 1996 launched over a dozen military operations, including strikes against a military housing complex, five attacks on naval targets, and ten on public security forces. Its targets, then, were clearly military in nature, in contrast to the broader assaults launched by Colombia's FARC and ELN.

With regard to its epistemological foundations, the group emphasized its devotion to "the scientific theory of revolution" and the significance of knowing "the scientific laws that govern the development of society and nature." It celebrated "the scientific method of analysis and interpretation of reality."[42] Earlier it was observed that the EPR attempted to reach the 'masses' rather than the more specific targets associated with identity politics, and that it depicted a static Mexico in which the exacerbation of traditional problems have plagued the country since the Mexican Revolution. In short, the EPR has adhered to a decidedly modern epistemology through which it interpreted classic Marxism-Leninism. Rather than acknowledging many truths, perceiving shades of gray, and constructing rainbow coalitions commensurate with identity politics, the EPR clung to a traditional framework. It did not acknowledge transformations associated with postmodernity and the RMA. Although at times the rebels belatedly and awkwardly attempted to stretch beyond their modern lenses, they have generally focused upon traditional problems and traditional-modern approaches to resolving them.

During their inaugural year, the rebels launched a total of 36 military operations. Meanwhile, the government was busy concocting a response that took a number of points into consideration. Chief among these was the potentially delicate nature of combating the EPR. This was due to complications stemming from the newfound context of Nafta, and from the increasing gaze of Northern NGOs and the media upon Mexico in the wake of the EZLN's appearance in 1994. The government also had to consider the potential global response to perceived instability in Mexico, especially in light of the disastrous peso crisis of 1994–1995.[43] For the Mexican government, then, the crux of the problem was how to neutralize the power of the EPR in a manner that went unnoticed by outsiders. This job was rendered easier by the rebels' traditional and outdated approach and by their incapacity to manipulate the global communication system. This meant that the military could get away with implementing fairly high levels of repression without attracting much national or global attention. The government is estimated to have placed about 40,000 Mexican troops in Guerrero between January and June 1997.[44] The government's strategy, according to the director of Guerrero's excellent human rights agency Tlachinollan, was to "create fear and division among the people, and to give the impression to those outside the region that everything is OK."[45]

Based on escalating infiltration of the EPR by intelligences services, the government had a rough idea of where to apply military pressure. But it frequently erred in its broad application of repression, and at times the state essentially terrorized remote and often indigenous communities. The rebels claimed that, in 1997 alone, the government kidnapped and tortured 200 people in the state of Guerrero on grounds of association with the guerrillas.[46] There were frequent military attacks such as the one on 6 April 1997 in San Miguel de Ahuelican, an indigenous town. Around 300 military troops in helicopters and jeeps suddenly arrived and swept the town looking for arms, and detained and tortured six men in an effort to obtain information on the insurgents.[47] Between August 1996 and February 1997, the military and police had arrested 70 people on charges of collaborating with the EPR. In a notable episode, one person claimed that he was "persuaded" and "tortured" by Mexican authorities, with two U.S. agents present, to increase his role from collaborating with the guerrillas to infiltrating them.[48] Based on intelligence efforts, a key EPR leader, Benigno Guzmán Martínez, was arrested in January 1997.

Two other phenomena are worth noting with respect to the EPR during 1997. First, in addition to the intensification of government repression, right-wing paramilitary squads began to (re)form to rout out guerrillas and suspected sympathizers. Their role was to generate a climate of fear that would intimidate *campesinos* from expressing loyalty to the rebels. Four paramilitary groups existed in Guerrero by 1997, including the Familia Peñaloza in Atoyac (which had existed since the days of Cabañas and which was revitalized during this era), another directed by Epifanio Hernández Vélez, one led by Bautista Catalán, and another by the Familia Flores. Second, in a sort of anti-intelligence campaign, the EPR concocted rumors of multiple guerrilla groups in Guerrero, with seven of these noted by the Mexican press by 1997.[49] Ironically, while the EPR invented this rumor as a tactical measure to deflect military aggression against itself, the group would actually fragment into a number of feuding organizations beginning in 1998.

Phase Two: Fragmentation and Debilitation, 1998–2000

In the Colombian case, we noted that although politics in general is fragmented, guerrillas remained unified overall. By contrast, in Mexico, politics tend to be centered, but the guerrillas under consideration tended to be fragmented to the extreme. For example, on 8 January 1998, the Ejército Revolucionario del Pueblo Insurgente (EPRI) was established by disgruntled members of the EPR. Given that the EPR lied about the nature of its original composition, should we consider reports of the emergence of EPRI as well as those of other insurgent groups to represent falsehoods crafted by the rebels to scatter the attention of authorities? The evidence suggests otherwise. Although EPRI indicated in its debut interview that it shared the same ideological framework as the EPR, it severely criticized the EPR's faulty "methodology of constructing political and military" strategies, and was especially doubtful regarding the "prolonged war" approach favored by them. The EPRI also indicated that it wished to expand its influence beyond the EPR's limited domain of Guerrero and Oaxaca.[50] Further, it bitterly accused the EPR of being dogmatic, too centralized, and antidemocratic. Overall, the creation of EPRI signaled frustration among some EPR loyalists regarding the lack of tangible progress.

The EPRI stepped into full public view with the El Charco massacre of 7 June 1998. One of the most balanced and distanced portrayals of the tragedy was offered by the Liga Mexicana por la Defensa de los Derechos Humanos, which released its report exactly one year after the massacre. What is clear is that the Mexican military suspected that EPR members were using a school in Ayutla de los Libres, Guerrero, as a safe house. Scores of indigenous community members had been at the school to attend a meeting on agricultural issues, with the discussion creeping on into early morning hours. Eventually, many of the attendees, including families, simply went to sleep at the school rather than going home, according to the testimony of the community's commissioner. An indigenous girl testified to the Liga that

> [w]e were sleeping in the early morning when the military arrived. They surrounded the school and started shouting, but I didn't understand because I don't speak Spanish well, only Mixteco, and I didn't understand what they were saying and then they started shooting and shooting.[51]

The massacre left 11 dead, including four EPRI members and seven innocent bystanders. Five others were badly hurt, and 22 were detained. One of those detained and later freed was Ericka Zamora Pardo, a young student who testified to the Liga that she had signed a confession of being linked to "terrorism and rebellion," but had done so after being sexually tortured.[52] While the tragedy of El Charco rightfully received much attention by the press, many smaller but equally repressive episodes did not. By the end of 1998, the government had 90 persons in captivity for being associated with the EPR or EPRI. By that time the rebels had killed at least 70 people, mostly members of the military or police.[53]

According to the ex-director of Mexico's federal intelligence agency, EPRI by the fall of 1998 was estimated to have approximately $1 million in wealth, with an operating budget of between $25,000 and $30,000 per month. As former members of the EPR and of its predecessor, PROCUP, the EPRI members were believed to have benefited financially from previous kidnappings of wealthy Mexican businesspeople, including Alfredo Harp Helú, Jorge Sekiguchi, and Angel Losada.[54] Given their frustration with the slow progress and limited geographical range of the EPR, EPRI's six columns of guerrillas apparently had plans to spread quickly into other regions of Mexico by the year 2000, including Mexico City, the state of Mexico, Morelos, Coahuila, Tamaulipas, Chihuahua, and Nuevo León. Nonindigenous leaders dominated the group, though it had support from key aboriginal areas of the La Montaña region of Guerrero, where Mixtecos and Nahuas predominated.

In an October 1999 interview with the popular Mexican political magazine *Proceso,* EPRI's Comandante Antonio suggested that Mexico was in a "pre-revolutionary period." EPRI wished to fill the void left by the EZLN, which he chastised for its military "deficiency" and for its failure to achieve a national presence. Further, Antonio had little faith in the democratic Left such as the PRD, which in his estimation had peaked in 1988. For EPRI, the traditional democratic route was not a viable one: "To conquer democracy" an armed democratic revolutionary struggle was necessary.[55] Like the EPR, EPRI adopted a traditional and modern approach, emphasizing class struggle as well as an armed insurrection to "conquer" the state. But EPRI all but fell apart days after that interview with the October 1999 capture of key EPRI leaders in a Chilpancingo safe house. Various essential documents of the group were confiscated by authorities. The military operation, based on intelligence from Centro de Información de Seguridad Nacional (CISEN), yielded three of the group's key leaders—Carlos García Rosales (Comandante Antonio), Gloria Arenas Arial (Colonel Aurora), and Ofelia Flores Nava.

Although EPRI and the EPR both referred to the "masses" and to the sweeping allegiance of class, the guerrillas themselves failed to embrace the same centralizing tendencies entailed in their prescriptions to society. Instead, the rebels spun centrifugally, peeling off new splinter groups on a regular basis. After EPRI's emergence, for example, the Ejército Villista Revolucionario del Pueblo (EVRP) appeared in 1999, which subsequently bombed a police station and various U.S. businesses including outlets of McDonalds and General Motors. While these isolated attacks grabbed headlines, the fractured and disparate nature of guerrilla activity rendered them steadily weaker from 1999 onward.

Phase Three: Fragmentation, 9/11, and the Withering of Traditional Guerrillas

Fragmentation continued apace in 2000, with the appearance of the Fuerzas Armadas Revolucionarias del Pueblo (FARP), which the director general of the Policía Preventiva identified as a "cell" that split from the EPR. The FARP was responsible for bombings in Puebla in February of that year, among other similar attacks, and also appeared in a Mexico City park on 8 April 2000, where it read a long diatribe beginning with references to Zapata. None of the group's remarks were ideologically or epistemologically distinct from those of the EPR.[56] The

FARP should be viewed as another loose cannon that frayed from the core. Also that year the FARP, the EVRP, and another group, the Comité Clandestino Revolucionario de los Pobres-Comando Justiciero 28 de Junio, occasionally joined forces to form the Coordinadora Guerrillera Nacional Jose Maria Morelos y Pavón. Like the EPR, their armed attacks were launched against the military or police, or consisted of bombings of corporate entities that were calculated to avoid the killing of civilians—something that differentiates them from Colombian guerrillas. Also worth noting regarding these groups was their tendency to stretch beyond the EPR's stronghold of Guerrero.

The EPR perpetuated its standard attacks, exemplified by a February 2000 assault on Mexican army soldiers in Coyuca de Catalán, and the kidnapping and robbing of a wealthy Guerrero rancher in December of that year. Further, the EPR criticized the new government of President Fox, the first president since the revolution who did not belong to the PRI. The rebels described his policies as a "continuation of the policies of Salinas,"[57] and lambasted his plans for privatization, for the taxation of food and medicine, and for what they described as his attempts to "Americanize" Mexico.[58]

At this point, the rebels recognized an array of strategic deficiencies on their part, but vowed to continue armed struggle. In July 2000, the EPR suggested that its "principal error . . . was to incorporate all zones of confrontation," a strategy that was based upon an "over-evaluation of our forces." The rebels also confessed that they moved too quickly from "irregular units" to an attempt at creating a "regular army." This is interesting, as it defies one of the most potentially fruitful elements of the RMA—the advantage of fighting as an irregular force. Despite key strategic and tactical errors, they restated their firm commitment to "scientific" Leninist theory.[59] Similarly, the EPRI, badly broken after the capture of key leaders during the previous year, indicated in April 2000 that the group would continue its fight and would not negotiate: "Dialogue for what?. . . . A dialogue should be based on mutual respect and a balance of forces, at this moment the state has an evident advantage over the armed insurgency."[60] While the rebels were aware that they were losing their contest with the government, they did not seem to understand the nature of this loss. What is striking is that the only major strategic errors of which the EPR/EPRI were conscious concerned excessive ambitions and timing. They did not recognize that their entire approach was based on the epistemology and ideology of a bygone era. If they had, they would have noticed more fundamental organizational problems. Beyond the assortment of deficiencies noted above, the rebels could attract the loyalty of only a smattering of local *campesinos,* rather than the strategically significant national and global support they might have received had they couched their traditional and often legitimate demands in a forward-looking strategy.

The year 2001 began with a familiar phenomenon, although at a slower pace. A series of three bombs were detonated at banks in Mexico City, where authorities first blamed the FARP and later assigned responsibility to the EPR. Rebel activity clearly fell to new lows, as the government escalated its repression especially in Guerrero and Oaxaca. In the case of the latter, by the beginning of 2001 at least 101 indigenous members of the community of Loxicha remained in detention by military authorities for alleged ties to the EPR. After the pivotal events of

11 September 2001, the EPR launched some small attacks in Tecpan and Atoyac, but clearly remained in low gear.

In 2003, EPRI members sent a communiqué to a local newspaper indicating that the government wished to "criminalize" them by falsely linking them to kidnapping, narcotrafficking, and other crimes.[61] Further, while suspected EPR members for years had been accused of terror and rebellion by the government, the term "terror" assumed a far more sinister connotation after 9/11. Now the EPR and its offshoots were charged with a crime that seemed to equate them with al-Qaeda. Clearly, the EPR and its tactics were distinct from that group and even from Colombian guerrillas, because the violence propagated by the EPR was never aimed at innocent civilians. Yet the armed attacks they committed were portrayed by the Mexican government to represent the link between terror, crime, and war. In addition to the weaknesses the EPR already suffered—in terms of strategic errors, fracture, and successful government repression—the new strategic context surrounding the discourse on terror led the EPR to rethink its tactics and this likely contributed to the waning of its armed attacks after 9/11.

By 2003, the EPR had disintegrated into at least half a dozen groups, some of which had further subdivided into cells.[62] Although the EPR and its offshoots were not skilled at using the media as the Zapatistas have been, they attempted to turn to their advantage the rising anti-Americanism linked to the U.S. invasion of Iraq. In a communiqué signed with their usual slogan of "win or die," the rebels criticized the United States for "imperial expansion" and for "genocide and crimes against humanity." The rebels pointed to what they viewed as symmetry between U.S. "imperialism" in Iraq and in Mexico.[63] Although the same kinds of criticism would provide the democratic Left throughout Latin America with fresh fodder to accelerate the decline of U.S. influence in the region, the EPR was not able to mimic that trend because the group attracted such scant public attention.

In 2004, authorities at CISEN characterized the EPR as "fractured, traditional, violent and small."[64] From the government's perspective, the EPR was a concern, but one that was growing weaker and that was clearly under control. Other experts in the area, however, were cautious at the time not to underestimate the power of the rebels. A prominent reporter in Guerrero suggested that the group's numbers were well above the 500–1,000 estimate by the government.[65] Similarly, a representative of a Guerrero human rights NGO located in a region where the rebels had been active argued that the "silence," or dearth of public activity, on the part of the EPR was a "tactic, not a sign of weakness."[66] The debate over whether or not the EPR was playing possum was settled the following year.

In April 2005, the EPR claimed that "[e]very day we have reached a better accumulation of force," but that "[w]e will realize no military action because we respect the time of struggle of the people."[67] This was a clear sign of the fundamental weakness of the group. As we have observed, the EPR never cultivated any significant ideological or political power. Now it found itself in a position where it was strategically unwise to rely on the only instrument of power it knew how to use—force. Despite the tone of military bravado in the rebels' statement above, their capacity to exert combative force was severely diminished both by government operations and by the EPR's own implosion. At the campus of the Universidad Autónoma de

Guerrero, traditionally a hotbed of leftist support, students and faculty suggested in spring 2005 that the EPR had been "broken."[68]

The slide continued with the release in September 2005 of "A Little More History," wherein the EPR revealed that it had lied about its origins, as noted earlier. Setting the tone of failure, the introductory paragraph contained an admission that the group had made a number of grave strategic errors over the past decade. The rebels underscored their awareness that they had not achieved the status worthy of a media "rating" or sound bite, and that the group had not been adept at using "theatrics"—a clear and jealous reference to their incapacity to attract the degree of national and global attention garnered by the Zapatistas. They lamented that early overtures the group had made toward the Zapatistas were received frostily. The EZLN and the Mexican Left in general, said the EPR, "satanized us as 'ultras.' "[69] More recently, there have been civil though distant relations between these distinct rebel organizations. The EPR dismissively referred to its offshoots, such as EPRI, as "deserters" who were "satisfying personal necessities"—suggesting that they relied on crime for self-interested purposes.[70]

At the end of 2005, the group released a fresh document that appreciated, among other things, the altered context of global politics. In its magazine, *El Insurgente,* the EPR noticed that U.S. power was receding in Latin America, as marked by Washington's failure earlier in the year to achieve a regional free trade agreement. The rebels also pointed to the popular backlash against newfangled neoliberal schemes such as the La Parota Dam project and Plan Puebla Panama (PPP). Further, as a group that had been badly hit by the efforts of the Mexican intelligence system, the EPR naturally attacked President Fox's plans to intensify the intelligence agency CISEN.[71]

The EPR—Analytical Conclusions

The EPR embraced a traditional ideological perspective that emphasized class struggle among Mexican *campesinos,* in alignment with workers and university students, to achieve the ideals that the Mexican Revolution defined. The most accurate range of its armed supporters was about 1,000–2,000 members at its peak in the late 1990s. The vast majority of these were *campesinos*—the mainstay of the Mexican Revolution—who had suffered mounting hardships in the country's agricultural sector beginning in the 1960s, problems that were exacerbated in the wake of Nafta after 1994.[72] Ideologically, these insurgent forces, which were centered in Guerrero but which had at certain points established a presence in several states of southern Mexico, borrowed strongly from classic Marxism-Leninism and from the ideals of Zapata and Villa. The rebels updated these with Guevara's guerrilla warfare tactics that were popular at the time of Cabañas' struggle, and adhered to Che's view that "[g]uerrilla warfare is used by the side which is supported by a majority but which possesses a much smaller number of arms for use in defense against oppression."[73] Although they heeded that advice on what today would be termed "asymmetric warfare," the EPR and its offshoots ignored a crucial element of Guevara's thought—his insistence on the importance of political development and education.[74]

When the EPR emerged in 1996, a year-and-a-half after the Zapatista uprising, it looked at first as if the EPR shared the EZLN's media-savvy and postmodern approach. Guerrero's debutant rebels initially appeared in new and neatly pressed uniforms for a fleeting photo op. But their inaugural bombing spree and other acts of violence quickly distanced them from the militarily bankrupt but ideologically powerful EZLN. Strategically, the EPR from 1996 to 1999 attempted to defeat the Mexican army in order to break the power of the bourgeoisie, and used tactics designed to target the military and other security apparatus. The group's strategy was altered in 2000, when it admitted that it had overestimated its military capacity and had not made the planned jump from an irregular to a regular insurgent army capable of standing up to the Mexican military. This in itself represented a significant strategic error, and betrayed an ignorance of a key element of the RMA—the potential power of irregular forces in contests with traditional state armies. The rebels' strategy was scaled down again in 2005, when the best the group could say was that it was "still present." By this point, its strategy was reduced to a desperate attempt just to exist.

The EPR and its offshoots failed to come close to achieving the goals they had defined for themselves, which is the fairest test of the group's power. We will recall that their initial objectives included the defeat of the Mexican military and the subsequent construction of a democratic socialist state aimed at meeting the material needs of Mexicans in general and *campesinos* in particular. What led to the EPR's failure? They did not have the benefit, as the Zapatistas did, of a strongly organized community as exemplified by Samuel Ruiz' nontraditional Catholic Church. That is, the EPR attempted to draw support from a society that was notoriously disparate and disorganized. Further, the group was clandestine, and could not openly use public spaces to press its political agenda. The closed nature of the group was a result of its reliance on armed force, which rendered it not only as an insurgent group but also as a terrorist organization in the eyes of the Mexican government. Their employment of violence was also repugnant to Mexican society in general, which had grown accustomed to the predominance of peaceful politics and defanged guerrillas in the form of the Zapatistas. Moreover, pronounced fragmentation and infighting signified that the group was imploding as a result of its strategic and political failures. Overall, their use of force diminished the potential for significant national or global support, as did the EPR's outdated ideological position, its stale epistemological framework, and its general failure to embrace elements of the RMA.

* * *

In the previous section we observed that one way to grapple with severe economic and political difficulties is to "fight," through the formation of social resistance groups, ones that range from armed insurgent forces analyzed in this chapter to peaceful social movements examined in the next chapter. Another way to cope with such despair is to take "flight," that is, to leave the country and seek illegal work in the United States, an entrenched arrangement that has served the interests of impoverished Mexicans as well as of American businesses that depend on migrants to remain competitive. This issue is perhaps the clearest example of complexity vis-à-vis Mexico and the RMA. It entails the shifting meaning of frontiers and competing conceptions of political space. The extent of this complexity is rooted in profound economic imperatives that rival newfangled conceptions of U.S. security concerns after 9/11. Let us proceed to unpack this manifestation of the RMA.

Complexity: Migration and Redefined Frontiers

> I don't know what to say to you, father. I don't even recognize you. . . . We're starving to death . . . all your descendants are about to kick the bucket and fall over dead. And what makes me so mad is that it's from hunger. Do you think that's fair and square?[75]
>
> *Juan Rulfo,* "Paso del Norte"

Unemployment and underemployment in Mexico, combined with the lure of higher wages in the United States, render illegal migration to the States an attractive and necessary option for many. Mexicans working illicitly in the United States sent home $24 billion in 2006, about a 20 percent increase over the same period the previous year. Remittances have been the largest source of legal foreign income for Mexico after oil. For example, in the first 11 months of 2005, remittances were valued at $18.3 billion compared to $25.7 billion worth of foreign petroleum sales during the same period.[76] Mexico is estimated to have received about 16 percent of total global remittances in 2005, compared to 31 percent for all of Latin America.[77] About 399,000 Mexicans are conservatively estimated to have migrated to the United States in 2005—this represents an eye-popping figure of more than 1,000 migrants illegally crossing the border on a daily basis.[78]

To reach a more vivid impression of what migration means to social forces in Mexico, let us focus on the case of Guerrero. About 31.6 percent of Guerrero's households received remittances, representing 8 percent of total household income in 2000 (the last year the Mexican government recorded such statistics prior to this writing).[79] Five percent of Mexico's total remittances—or about $915,000 during the period January-November 2005—was sent to Guerrero, where 3 percent of the country's population resides.[80] Those who have migrated from Guerrero typically have done so due to the crisis in the agricultural sector, the incapacity of tourism to absorb new entrants into the economy, and wages being too low for subsistence. These factors are common to Mexicans in the rest of the country who choose to migrate illegally to the United States. Guerrero ranks tenth among the country's 31 states in terms of percentage of migration to the United States.[81] Most migrants from Guerrero have been destined for Chicago, New York, Atlanta, or Los Angeles[82]—but their trek has become exponentially more perilous since 11 September 2001. While specifics regarding the tightening of border security will be discussed shortly, suffice it to say here that increased U.S. attention to the border has meant that Mexicans are pressured to cross the frontier in more dangerous locales, especially in mountainous regions or the desert. A record number of migrants died in their attempt to cross the border in 2005—464 people looking for work so they could send some money back home.[83]

The price migrants pay to coyotes or *polleros* (illegal smugglers) has risen steadily since 9/11, resting in early 2006 at about $1,500 for the trip from Guerrero to the United States, and at about $2,500 from the southern border town of Tapachula, Chiapas. The business of human smuggling in Mexico is estimated at about $300 million for 2003, and presumably has risen since that time as coyote fees have increased.[84] Perils appear not only at the border but also all along the trip. Directors of NGOs who work with migrants in Guerrero and Chiapas en route to the United States point to many such dangers. For example, Mexican border agents and police sometimes extort money from migrants at the southern border. In other cases, many migrants (especially children) are abandoned by coyotes after payment has been made, common migrant routes are littered with banditos who are aware of migrant traffic patterns, and many women and girls fall into the sex trade along the way. Migrant behavior and profiles have transformed against such a backdrop. After 9/11 they typically have planned longer stays in the United States due to the higher cost of human smuggling and the risk of failed reentry into the States after a periodic visit to their families in Mexico. Increasingly, only the young and hardy have attempted to make the trip to the United States.[85]

What is clear is that Mexican migrants are suffering key violations to human rights and security. In January 2006, Mexican presidential candidate Manuel López Obrador indicated that "the objective is that nobody should have to abandon the country and their family to find work and live in dignity."[86] But the reality is that many Mexicans must abandon their families just to subsist. Here there are multiple rings of breached security: The legitimate interest of the U.S. to have tighter border security after 9/11 is breached by more than 1,000 illicit Mexican migrants on a daily basis. And marginalized Mexicans—impoverished victims of a failed development model—risk all kinds of security threats, including death, just to eek out a living. A sharply slower rate of growth for remittances sent to Mexico in the first months of 2007 suggested that increased border security was beginning to limit the increasing flow of migrants into the United States, a phenomenon that in the past has meant greater desperation and danger for Mexicans illicitly seeking work in the States.[87] The strategic space of the border is complex to the extreme. Let us proceed to examine elements of refashioned border security.

Surveillance: The Case of a Redefined Frontier

> There's one frontier we only dare to cross at night. . . . The frontier of our differences with others, of our battles with ourselves.[88]
>
> *Carlos Fuentes,* The Old Gringo

Ultra-surveillance is a key component of the RMA. While the next chapter will explore other facets of this in relation to a variety of social movements, here we shall examine its ample expression with respect to illicit Mexican migration. Since 9/11, the context of border security has changed radically, prompting Washington to harness ultra-surveillance devices as a principal means for plugging strategic holes related to its War on Terror. As we shall see, the application of such instruments is very much a work in progress, one that has been marked by both a laggard pace and remarkable glitches in what was supposed to count as technological wizardry.

The crux of the problem from the U.S. perspective, as explained in 2005 by David Aguilar, chief of the U.S. Border Patrol, is that "[w]e are concerned that illegal human smuggling routes may be exploited by terrorists to conduct attacks against the US homeland. Reducing illegal migration across our borders may help in disrupting possible attempts by terrorists to enter our country."[89] But the task is monumental. We already observed that more than 1,000 illegal migrants from Mexico are estimated to cross the border on a daily basis. Further, about 85 percent of those caught and deported have returned to the United States within 24 hours.[90] Senator Edward Kennedy lamented in early 2006 that "[w]e have spent $20 billion on chains and fences . . . in the southern border over the last 10 years. It doesn't work."[91]

Although the U.S. government has spent 58 percent more on border security since 9/11, standing at $7.3 billion in 2006, and has reorganized bureaucratically to allow the Border and Transportation Security Division of Homeland Security to take over national borders and ports of entry beginning 1 March 2003, controversies concerning the glaring weaknesses of the U.S.-Mexican border have mounted. On paper, it looked as if Homeland Security was unleashing a system of ultra-surveillance that would, from an American perspective, render the frontier more secure. In his discussion of the "defense in depth" strategy of the border, the chief of U.S. Border Patrol told a congressional committee in 2005 that Homeland Security is committed to increasing "the use of technology including the expansion of camera systems, biometrics, sensors, air assets, and improving communications systems" that together can provide a "multiplier effect."[92]

The post-9/11 implementation of surveillance technology at the U.S.-Mexican frontier has been laden with difficulties. Testifying in June 2005 before the House Homeland Security Subcommittee on Management, Integration and Oversight, the U.S. deputy inspector general provided a dismal account of the status of the Border Patrol's Remote Video Surveillance program (RVS). The project involved the installation of surveillance cameras with special lenses capable of night vision, the construction of towers for microwave transmission equipment, the installation of monitoring equipment at Border Patrol headquarters, as well as the maintenance and repair of the equipment. The surveillance cameras were expected to do the work of four to five Border Patrol agents at any given time. The deputy inspector general cited "significant deficiencies" in the project, including: (1) the failure of surveillance devices to be installed in the promised timely fashion; (2) surveillance devices installed being "incomplete and unreliable;" (3) a lack of competition for contract awards; (4) the use of inappropriate contract vehicles; (5) inadequate contract administration and contract management; (6) overcharging for equipment and failure to provide promised equipment (e.g., thermal imaging cameras lacked the promised "doubler lens"); and (7) the failure to follow government guidelines regarding wages, among other problems.[93]

Members of Congress were stunned to hear tales of faulty wiring and of cameras that regularly "swivel out of control in hot weather," and generally condemned this case of "gross mismanagement of a multi-million-dollar-contract."[94]

While the RVS system on paper looked as though it would be a high-tech marvel featuring the power of the constant gaze entailed in Bentham's panopticon, by 2006 it proved to be a near comedic folly. Similarly, a representative from the Department of Homeland Security indicated that not only is the border as insecure as ever, but also the new bureaucracy itself is bloated and inefficient. This has been due to the restructuring entailed in the creation of the massive department, and its absorption of other bureaucracies such as the Border Patrol.[95]

* * *

A crucial and highly controversial U.S. project along the frontier is the $2.5-billion Americas Shield Initiative, which includes a fence along the entire border with Mexico, complete with cameras and sensors. Compounding this process of militarization at the border, President Bush announced in May 2006 that thousands of members of the National Guard would be deployed along the U.S.-Mexican frontier. All this conjures images of a Berlin Wall in reverse, because it is designed to keep people from entering rather than leaving. The plan received energetic support from some politicians who pander to local concerns in border states where binational drug-related violence and other matters have inflamed public opinion. The fence, which the Bush administration indicated in 2005 would be accompanied by drones, is problematic in a number of ways. Not the least of these is that the U.S. economy depends on the illegally cheap labor of millions of illicit Mexican migrants. This means that if such a fence were to be constructed, the current flow of illicit migrant traffic would have to be regulated by the government in a manner that satisfies their obvious economic demand in the United States. Conscious of their material value to the country, millions of Mexican migrants took to the streets in America in March 2006 and again on May Day of that year to protest against the threat of being criminalized under new legislation. The Bush administration responded that it was working on the monumental task of devising a program to institutionalize and regulate illicit migrant flows in a way that is compatible with a highly militarized and terror-proof border.[96]

Predictably, all of this evoked harsh criticism in Mexico. Former president Fox responded by observing that "[t]his situation we're seeing—a disgraceful and shameful moment where walls are being built, security systems are being reinforced, and human and labor rights are being violated more and more—won't protect the economy of the US."[97] The existing and promised networks of surveillance systems, walls, biometric inspections, vigilante groups, and so on, are likely to mean continued hardship, abuse, and even death for Mexicans seeking work in the United States. There have been more than 4,000 deaths of Mexican migrants seeking work there during the period 1994–2005,[98] a death toll that is likely to be augmented with an increasingly militarized frontier. In contrast to its aspirations for achieving first world status through Nafta, and for the establishment of a more seamless border, a rebuffed Mexico has found itself being sealed off from the United States.

These are not Mexico's only concerns. Critics have charged that the country is losing its sovereignty in the face of mounting U.S. efforts to pressure Mexico to bolster its northern and southern borders in the wake of 9/11. They pointed to programs such as Mexico Seguro, which was designed especially to combat drug-related violence emanating from Mexico. Respected Mexican intellectual Aldofo Gilly, for example, called the situation a "plan to militarize the northern border to suit the 9/11 interests of the US," and said that this amounted to an "abdication without precedent of national sovereignty."[99] Another point of contention concerned the reorientation of Mexico's intelligence services. Within the context of Washington's War on Terror, Mexico's chief intelligence agency, CISEN, has been expected to conduct more intelligence operations at the country's frontiers and ports of entry. But its budget has been cut by some 30 percent in 2004 alone, leaving it with what could be the insurmountable task of accomplishing far more with substantially less funding.[100]

Surveillance also has been bolstered on Mexico's southern border, where over 40,000 illegal migrants arrived in both 2004 and 2005—most with the hope of subsequently entering the United States. Central Americans have endured considerable hardship on their journey north. About 40 percent of all abuses in Mexico that are related to human trafficking occur in Chiapas near the border town of Tapachula.[101] It is doubtful that increased surveillance at the southern border would halt the flow of migrants, given their economic despair in combination with the mountainous jungles of the region that would prove difficult to monitor. The increased surveillance is more likely to increase the repression and abuse suffered on the part of those who are simply looking for work.

Overall, the migration issue is relevant to the RMA in at least three primary ways, all of which are interrelated. First, it entails instruments of ultra-surveillance in relation to security and strategy. Next, it raises for Mexico the strategic implications of adjusting its policies regarding borders, airports, and ports to the U.S. preoccupation with terrorism after 9/11. Third, it involves the redefinition of border security to cope with effects emanating from regional or transnational production under Nafta. Moreover, the extent of Mexican illicit migration to the States—with some 11 million such persons in the United States by early 2006—suggests profound economic integration between the two countries that goes far beyond what is legally enshrined in Nafta.

The migration issue, then, demonstrates numerous links between the realms of security and political economy. One such link combines social inequity with strategic themes. The U.S.-Mexican border, as a militarized boundary between a developing country and the world's richest nation, serves to reinforce social inequity. In effect, the border protects the rich from the poor.[102] It is a line of demarcation, not only between Mexico and the United States, but also in the net transfer of wealth between Latin America and the developed North to the tune of nearly $156 billion between 2001 and 2004 alone.[103]

Finally, Mexican illicit migration to the United States—representing the world's largest flow of illegal migrants—should be viewed as a form of displacement. By way of comparison with Colombia, geographic fate has provided Mexico's majority population with access to the safety valve of illegally working in the United States. No doubt this has bolstered Mexican political stability. This has not been the case with volatile Colombia. While that country's poor could conceivably make the daunting trip through Central America and then through Mexico, the grueling and extraordinarily perilous nature of such a trek has rendered it a highly impractical option. Further, in the Colombian case, we saw that displacement has been induced largely by warfare and has been both internal and external. In the Mexican case, as exemplified by the situation in Guerrero, displacement is economic in origin. While some are displaced to Mexico's large cities to work in the informal sector, the focus here has been on the security themes associated with the millions who illegally cross the frontier to work in the United States.

Conclusion

> I'm beginning to pay. The sooner I begin, the sooner I'll be through.[104]
>
> *Juan Rulfo,* Pedro Páramo

This chapter began with the observation that elements of political economy are key to Mexican security. During Cabañas' struggle in the 1960s and 1970s, the central problems were the exclusionary politics of the *cacique* system, economic inequity, as well as troubles emanating from a weakening agricultural sector and from the illicit harvesting of wood. These problems have continued to plague Mexico in places such as Guerrero. They have been compounded by other worries, such as increasing economic pressure to migrate north in search of illicit work, the social perils of privatization, a rising tide of violence and corruption linked to narcotrafficking, and so on. This has underpinned many aspects of social struggle, including insurgency.

Mexico's model of development has failed in some important respects. Most major sectors of the economy have experienced profound difficulty in the new millennium, including the manufacturing, agricultural, and tourism sectors. This has fueled an illegal exodus to the United States of over 1,000 people a day—leaving Mexico dependent both on their remittances and on the antirevolutionary "safety valve" of mass migration. Hosting some 11 million illicit Mexican migrants, the United States has become dependent on illegally cheap labor. Together with the legal flow of goods and services associated with Nafta, the enormous stream of illicit migrants suggests a profound blending of political and economic space between the two countries—a reality that has not been fully appreciated and articulated by the United States and that has escaped appropriate institutions and legislation.

These economic tendencies have been contradicted by another mandate that emerged after 9/11: the War on Terror. This has meant a push from Washington to seal its border with Mexico through the proposed creation of a militarized wall patrolled by high-tech surveillance devices and even by drones. Essentially, there has been a clash between models of political space that are mutually exclusive. On the one hand, economic forces suggest a larger and more integrated space of "North America." On the other hand, Washington's reaction to 9/11 points to a Fortress America whereby the United States is cordoned off by military barricades. The border, once referred to as a "scar" by the celebrated author Carlos Fuentes,[105] has transformed into a mirror image of the Berlin Wall. Here, then, we confront key elements of the RMA in the form of rupture, changing notions of political space, ultra-surveillance, and complexity.

While the Mexican government has faced an important and complex contradiction between the integrative tendencies of neoliberalism and Nafta versus the isolation of Fortress America under the War on Terror, we have seen how the discourse on terror has served some chief objectives of the Mexican state vis-à-vis insurgencies in Guerrero and elsewhere in southern Mexico. The discourse on terror has framed the construction of good versus bad guerrillas in the country in a fashion that is starkly reminiscent of Nietzsche's classic, "Good and Evil, Good and Bad."[106] The EZLN has been portrayed by the government, which itself has used terror, as the "good" insurgents, largely due to their almost complete lack of military capacity. By contrast, the EPR and its multiple offshoots have been cast as "bad" due to their use of violence, which the state has deemed to represent "terror." According to the EPR, the discourse on terror created by the state was to some extent shared by other leftists who viewed them disdainfully as "ultras." They were "bad" because they resorted to violence, they were closed and isolated, they employed a totalizing class analysis left over from the 1960s, and so on. Despite their legitimate gripes, they were "bad" because they were modern, traditional guerrillas in contrast to the postmodern and postmilitary Zapatistas. Overall, then, the U.S. War on Terror was manipulated by the Mexican government in a manner that served the interests of both the U.S. and Mexican governments by helping them quash an emergent rebel group. This discourse, then, helped the government make good on the "never again" mantra born in the wake of the surprise emergence of the EZLN in Chiapas.

Social forces in Mexico have also faced the formidable challenge of attempting to shape elements of the RMA to work in their favor. We have focused on the dynamic of 'fight or flight', whereby some social forces have attempted to deal with an abysmal social and economic reality either through political struggle or through migration. These social forces reside in the peripheral space where Mexican hegemony was never strong and where it has steadily receded even further since the 1980s. This is especially the case with respect to southern Mexico, where there is a strong element of identity politics involving the indigenous, the *campesinos*, and the traditional quest for origin.

All this is illustrative of the "package deal" that defines the RMA—that is, it is the bundle of features of the RMA that work together for complex, synergetic, and sometimes even contradictory effects. For example, in the context of economic globalization, Mexico's integration with the United States accompanied the adoption of neoliberalism. This entailed the abandonment of "Revolutionary Nationalism" and the "quest for origin" that defined the traditional era for the false promise of insertion into the first world. In peripheral Mexico, where hegemony wore thin, insurgents and other social resistance movements emerged—some of these were versed in elements of the RMA and succeeded in some of their endeavors, such as the EZLN, and others did not, such as the EPR. This package also included Mexico's strategic alignment with the United States, and its sudden predicament of being thrust by Washington into the position of junior partner in the War on Terror. As a result of this, Mexico has forsaken its once prominent role as a major Latin American leader. Here, then, is a melting pot for a volatile recipe that features the forces of neoliberal globalization, asymmetry, complexity, surveillance, regional strategic alignment, and the War on Terror. This synergetic effect of the RMA will be explored further in the concluding chapter.

* * *

The themes discussed in this chapter suggest important distinctions and parallels between the Mexican and Colombian cases in relation to the RMA. With regard to peculiarities, Mexico's location next to the United States raises special circumstances found in no other Latin American country. Further, while crime is intimately related to insurgency in Colombia, this has not been the case with the militarily weaker Mexican guerrilla groups. While Colombian guerrillas and their penchant for violence have had an effect on the daily lives of most of the population, Mexican guerrillas such as the EZLN have had a national impact only in the realm of ideas, while violence-prone insurgents such as the EPR have had little effect on the nation as a whole. Much more broadly, the role of the state as well as political culture are profoundly different in the Mexican and Colombian cases, which means that the prism through which elements of the RMA are filtered often produce distinct manifestations in the two countries. This distinction rests largely on the former hegemony of the Mexican state, present from the 1930s until a period of deterioration during the late 1960s through 1982. While hegemony has receded in Mexico, some of its effects linger. This has meant a relatively strong central government, along with the absence of civil wars and of the crippling tendency toward nationwide violence and fiefdoms that have blighted Colombia. This theme will be elaborated upon in the concluding chapter, where the cases of Mexico and Colombia are more fully compared.

Despite these and other important distinctions, there are also some noteworthy parallels between Mexico and Colombia. Rising social inequity in both countries has yielded crucial strategic problems, as has the trend toward privatization. They are among the very few Latin American countries to host significant insurgencies, with these concentrated in areas facing agricultural crisis, political exclusion, and victimization of the population by what should count as legitimate government security forces. Further, mass displacement represents a prominent strategic concern in Colombia, where this tends to be internal in nature, and in Mexico, where it is expressed through illicit migration. Finally, and most important for our purposes, a similar array of features of the RMA is apparent in both countries. Within the framework of "fight or flight," this chapter has focused upon terror, asymmetry, complexity, and surveillance. Privatized war and public resistance is the central theme of the next chapter's analysis of Mexican social forces and the RMA.

7

Economic Dimensions of the RMA

Privatized War, Neoliberal Strategy, and Public Resistance

Chapter 5 examined the struggle of Lucio Cabañas and his compatriots, who were met with a Dirty War that marked what was perhaps the most disturbing epoch of Mexican postrevolutionary history. Cabañas and his supporters struggled against the authoritarian and exclusionary *cacique* system, and against an agricultural crisis that has grown steadily worse ever since. The government's extraordinary campaign of murder and repression silenced the guerrilla movement of that era. That sad episode demonstrated the collapse of the PRI's hegemony, as it resorted to brutal coercion when it could no longer garner consent from peasants who were once exalted as the heart of the revolution.

Chapter 6 focused on the dynamic of fight or flight within the broader context of neoliberal economic restructuring that commenced in the 1980s. Economic factors have created stress for social forces, and this has promoted related strategic concerns. Important but underresearched Mexican insurgents were highlighted, especially those in Guerrero and Oaxaca that fell off the radar screen of Northern observers who preferred to focus on the ideationally potent but militarily powerless Zapatista movement in Chiapas. The "fight" launched by insurgents has been twinned with the dangerous and illicit "flight" on a daily basis of almost 1,000 Mexicans desperately seeking work in the United States. This phenomenon, which has represented nothing short of an economic necessity for both the United States and Mexico, has clashed with post-9/11 conceptions of border security. With regard to the RMA, the key features observed within the context of fight or flight are asymmetry, insurgency, terror, complexity, and migration.

Building on these discussions, this chapter examines two related spectrums of security concerns: from economic security to human security, and from privatized war to public resistance. With regard to dimensions of economic security, the strategic considerations associated with both the illicit and legitimate economies will be analyzed. This includes the "big ticket" areas of narcotrafficking in the criminal arena, and with regard to the legal economy, the focus shall be on security linked to energy production and the creation of new economic spaces. The discussion will then shift to human security and public resistance. In contrast to the last chapter, which concentrated on guerrilla activity, this one will focus on the peaceful struggle waged by social forces in relation to security. There will be a particular emphasis on the challenges they have faced utilizing surveillance in efforts to shine the light of the global gaze upon their various struggles. Overall, the spotlight will be upon elements of the RMA including the privatization of warfare, energy politics, identity politics, and biopolitics.

Dimensions of Economic Security: The "Illicit" Economy

Narcotrafficking: The RMA, Illicit Consumption, and the Erosion of Security

It is important to note at the outset that the Mexican narcotrafficking industry is not new. It has been present in Mexico since at least the 1930s, when opium cultivation developed to serve the U.S. heroin market. Washington later encouraged Mexico to cultivate legal opium to treat those wounded in World War II, and in so doing unwittingly bolstered the industry's capacity to produce heroin for American street use. Beyond opiates, marijuana production in Mexico intensified in the 1960s and into the 1970s to meet the Northern demand created by the hippie generation.

A major boost in the illicit drug industry occurred in the 1980s within the context of the debt crisis, when pronounced economic duress in Mexico during this period meant a welcome mat for the badly needed dollars that flowed from fueling the habits of American drug users. For example, Mexican opiates destined for the United States were estimated at about two tons in 1984, and rose 300 percent by 1988. The illicit trade was boosted further in the 1990s, when the U.S.-Mexican border became more porous to facilitate rising tides of trade linked to Nafta. Since that time, as we shall see, the Mexican illicit drug industry has boomed and in many ways has fused with Colombian and other Andean interests.[1] Until 9/11, perhaps the biggest security threat emanating from Mexico was narcotrafficking and related crime.[2] Despite the preeminence of the War on Terror, it continues to represent a huge security crisis for both the United States and Mexico.

The U.S. Drug Enforcement Agency estimated that 92 percent of cocaine arriving in the country passed through Mexico in 2004, compared to about 77 percent in 2003. This suggests ever stronger bonds between Colombian and Mexican cartels.[3] Mexican narcotraffickers are thought to have learned from their astute and highly entrepreneurial Colombian mentors a host of organizational, managerial, and strategic lessons. These include a heavy investment in counterintelligence, splitting into a multitude of smaller syndicates or cartels that are less conspicuous, diversifying wealth, strategies of money laundering, and so on.[4] Further, there is mounting evidence that Mexican traffickers have moved beyond a strict reliance on their Colombian cocaine connections to establish fresh contacts in Peru's Huallaga Valley. Mexican cartels, then, appear to be every bit as entrepreneurial and growth oriented as their notorious Colombian colleagues. More broadly, money laundered by narcotraffickers is estimated by the Inter-American Development Bank to represent 6 percent of Latin America's GDP in 2005, or about $120 billion.[5]

About two-thirds of all Mexican poppy cultivation is estimated to occur in Guerrero,[6] in addition to its title of third place in the country for marijuana production. The port of Acapulco and the coastline of Guerrero are thought to be major transit points for Colombian and Peruvian cocaine en route to the United States, as well as the site of arms purchases by Colombian cartels and insurgents.[7] A noticeable escalation of violence in Acapulco became obvious since late 2003 and especially in the beginning of 2004, when a dozen people were murdered due to narco-violence in the first three months of that year. Scores of murders have been reported in the local press since that time, with a pronounced increase in both the pace and ferocity of what increasingly looks like narco-warfare.[8] By mid-2005, illicit drug violence became virtually a daily phenomenon. Much of this has been attributed to the Golfo Cartel's placement in Acapulco of about 125 "zetas," or paramilitary hit men who serve as the security structure for the cartel, to wrestle strategic turf from rivals.[9] Between 2005 and 2007, wars among Mexican drug cartels in tandem with an intensification of the trade resulted in pronounced security problems that sometimes morphed into sheer terror.

In February 2006, the attorney general of Guerrero called for a narco-violence alert due to a wave of terror and mayhem unleashed by the feuding cartels of Sinoloa de Joaquin El Chapo Guzmán versus Golfo de Osiel Cárdenas.[10] Ultraviolence by narcotraffickers in early 2006 inspired the panicked declaration of a "circle of security," or the establishment of a shared preventative wall of dubious utility, by Acapulco and four neighboring communities—San Marcos, Tecoanapa, Tierra Colorada, and Coyuca de Benítez. As the two cartels competed for turf in the important Pacific port, public spectacles of ultraviolence seemed to be the tactic of choice. For instance, three grenades were hurled at the house of the state's public security director on 10 February 2006, and 20 Mexican soldiers were ordered to Acapulco in January of that year after a shoot-out on the streets of the city between police and cartel members that left four people dead. An escalating wave of armed attacks by narcotraffickers in 2007—including bombings, public assassinations, and kidnappings—roughly resembles the warfare tactics launched by the quintessential Colombian cartel leader Pablo Escobar in the late 1980s and early 1990s. Thus, Mexico has faced warfare emanating from both guerrillas and narcotraffickers.

Much of the alarming violence, terror, and warfare breeding in Guerrero has been ignored by the American press, which predictably paid more attention to a similar phenomenon occurring along key border towns. About 500 people were estimated to have been murdered due to narcotics-related violence in American and Mexican frontier towns between 2004 and 2005.[11] During the first seven months of 2005, 800 people were killed throughout Mexico due to violence associated with illicit drugs.[12] About 2,000 people were murdered in drug-related violence during 2006, and about 500 were killed in the first 11 weeks of 2007.[13]

Related extravaganzas near the U.S.-Mexican border were plentiful during this period. One of these featured several men masquerading as Mexican military personnel entering the United States during a smuggling operation,[14] and another involved the discovery in January 2006 of the longest and perhaps most sophisticated drug smuggling tunnel ever to undermine the border.[15] Adding to a sense of flashing terror has been the murder of at least four journalists during 2000–2006 due to their coverage of narcotrafficking, which may have limited reporting of the issue during that period. A total of 31 reporters were killed in Mexico during this period.[16] Joel Simon, of the Committee to Protect Journalists, said, "The situation (in Mexico) is very comparable to Colombia in terms of self-censorship and the level of violence."[17]

A crucial dimension of narcotrafficking in relation to the RMA is the intense force of corruption to which politicians, the military, the police, and other security structures are vulnerable. Examples abound of this tawdry relationship between crime and deteriorating security. A key aid to former president Fox, who had access to his every move for a period of three years, was discovered in 2005 to be a spy for a drug cartel, according to Mexico's attorney general Rafael Macedo de la Concha.[18] There are innumerable examples of corruption of military forces by narcotraffickers. It is important to emphasize that this is not to brandish the entire security establishment with charges of corruption, but only to point out that key elements have fallen prey to it. In 2003, for example, an army deserter testified that the military protected cocaine flights near Mexicali for the Arellano Félix Cartel between December 1996 and January 1997, and that military heads received $450,000 each for their services.[19] Members of many NGOs and civil society groups in Guerrero have observed that the military is present in areas known for opium cultivation and appear to protect such operations rather than combat them.[20] Beyond elements of the political sector and the military, police forces also have been leased by narcotraffickers. For example, two "sicarios" or hit men captured from the Chapo cartel in 2006 were later revealed to be policemen from Guerrero.[21] These are just a few examples that illustrate the hugely corruptive influence of this industry. While we noted in the earlier chapters on Colombia that the United States is not immune to this sort of corruption, its influence may be more extensive in Latin America given the relatively low pay received by members of the military, police, and so on.

There is an assortment of other noteworthy security issues associated with narcotrafficking. First, it is closely tied to the illicit arms trade. Much of these weapons come from the United States, where guns are legal, and flood into Mexico, where they are not. Further, Mexico's attorney general's office indicated in June 2005 that they had evidence of drug cartels purchasing arms over the Internet.[22] Second, financial institutions also have been implicated in corruption and other crimes related to narcotrafficking. In January 2006, for example, U.S. authorities alerted Mexico's attorney general's office of likely collusion between functionaries of the Banco de México and the Juárez Cartel. Presumably, that is the proverbial tip of the iceberg, given the enormous amounts of money laundered by narcotraffickers. Third, narcotrafficking has ecological implications. Mexico's National Commission for Forests estimated in 2003 that 20 percent of the forest fires in the Costa Grande region of Guerrero were due to narcotics cultivators burning forests to plant crops—hence the issues of uncontrollable forest fires, deforestation, erosion, loss of wildlife, smoke pollution, and so on.[23] Finally, and crucially, the "war on drugs" has long been an excuse to militarize regions in an attempt to combat what is perceived as subversive activity—a point that will be revisited.[24]

Overall, given that the drug trade is driven more by Northern demand than by Southern supply, it presumably would have existed even if Mexico's legitimate economy had not been as debilitated as it has been for much of the post-1982 period. But the weakness in the Mexican economy certainly has catalyzed narcotrafficking, since it fills gaps in employment and pads bank coffers. It can also yield lucrative spin-off effects in the legitimate economy when narco-dollars are reinvested in sectors such as high-end construction and luxury tourism. Just as often, however, it can lead to the distortion or deterioration of the legitimate economy. With regard to distortions, legitimate entrepreneurs may have difficulty competing with huge investments made by narcotraffickers that are designed to launder money. Narco-violence has had a negative impact on the tourism industry. In November 2006, for example, the Canadian-based Conquest Cruise canceled tours to Guerrero during the 2006–2007 high season owing to the escalating narco-violence in the state. This has created panic in Acapulco's tourist sector.

There are some important parallels as well as some salient distinctions between the strategic implications of the drug trade in Mexico and Colombia. In terms of similarities, in both countries narcotrafficking has spawned corruption, violence, and ecocide, and has contributed to other crimes such as the illicit arms trade and money laundering. These countries differ substantially with regard to the relation between illicit drugs and insurgency. In Colombia, the drug trade underwrites insurgent groups to the tune of millions of dollars annually. The FARC and the AUC, for example, have become formidable military machines due to their relation to illicit drugs. In Mexico, guerrilla groups such as the ELZN have had no discernable relationship at all to narcotrafficking. Others, such as Guerrero's EPR or EPRI, are alleged to have had occasional and tenuous ties to the industry, but surely it has not been their empowering "money tree" as it has been for their Colombian compatriots. In Colombia, then, narcotrafficking and guerrilla warfare have been intertwined. In Mexico, insurgents have been distinct from drug runners. But illicit drug trafficking in Mexico is related to serious narco-warfare that can amount to urban terror. It is worth underscoring that narco-warfare has produced far more murder and violence in Mexico than have guerrilla insurgencies. Drug-related warfare has been apparent not only in Guerrero but also in key points near the U.S.-Mexican frontier, in Mexico City, and in scattered locations across the country.

The RMA and Other Elements of the Illicit Economy

While narcotrafficking is by far the largest element of illicit economy in Mexico and in Guerrero, other types of crime are also rampant and are related to security themes associated with the RMA. Illegal wood harvesting is particularly noteworthy in

Guerrero, where in 2005 the director of Greenpeace Mexico noted that 70 percent of the wood harvested in the state was illegal.[25] Typically, in Guerrero, deforestation is followed by the planting of illicit crops such as opium poppies.[26] This has been corroborated by several members of NGOs and others I interviewed in the state.[27] There are a number of security issues surrounding this. Massive illicit harvesting of timber in mountainous areas creates ecological devastation in terms of erosion and sometimes flooding at lower elevations. Organized *campesino* protestors have often met with peril and intimidation at the hands of paramilitary forces working for *caciques* that dominate such operations. The *campesinos* have also been victimized by the government, which has imprisoned key leaders of protest movements by framing them for crimes they did not commit.

There are at least two sets of especially notable cases in this regard. The first involved the arrest of Rodolfo Montiel and Teodora Cabrera, members of the Organización de Campesinos Ecologistas de la Sierra de Petatlán y Coyuca (OCESPC), in 1999. They led a group that blocked roads to Boise Cascade's clear-cutting site, which they argued was supported by a network involving *caciques*, local leaders of the PRI, and paramilitary security forces. The two were charged with drug trafficking, after being tortured and forced by the police to sign confessions.[28] They were released on "humanitarian grounds" almost three years later, in November 2001, under a decree by President Fox, who faced international condemnation following the unsolved murder of the ecologists' high-profile lawyer, Digna Ochoa, just a few weeks earlier. A second and similar case that has received considerable attention is the arrest of another OCESPC member in 2004, in tandem with arrest warrants for 14 other members of the group, for the alleged murder of the son of a logging boss. Many international human rights groups, including the Washington Office on Latin America, viewed the arrest as an attempt to intimidate other members of the group from protesting against illegal logging.[29]

Besides narcotrafficking and illegal logging, kidnapping has reached alarming proportions in Mexico, with the country in a neck and neck competition with Colombia for the world title of first place. The number of reported kidnappings in Mexico has increased from 200 in 1994 to about 1,300 in 2002. Many kidnappings go unreported for fear of injury to the victim and also because, according to President Fox's public security chief Alejandro Gertz Manero, many kidnappings are linked to the police.[30] In the first half of 2005, Mexico pushed ahead of Colombia to host the world's highest number of kidnappings—194 compared to 172 in Colombia.[31] Given that Mexico's population is almost three times that of Colombia's, Mexico's per capita kidnapping rate is lower than Colombia's, but nevertheless remains alarming.

Dimensions of Economic Security: The "Legitimate" Economy

We shall now turn to the two key themes in the legitimate economy that are closely related to the current RMA—the security implications of energy production and of the creation of new economic spaces—and discuss their strategic implications for Mexico. To begin, security issues related to petroleum extraction and the development of hydroelectric facilities will be examined. Also key to the RMA have been security concerns related to the new spaces created by globalized and regionalized production. With this in mind, there will be a focus on various strategic aspects associated with the emerging neoliberal trade area embodied by Plan Puebla Panama.

The RMA and Energy Production: The Petroleum Industry

The political economy of energy is closely related to the current RMA on a global scale. Though this has been most obvious in the Middle East since the Gulf War of the early 1990s, it has found ample expression in Mexico since 9/11. While the petroleum industry no doubt counts as a bright spot for the Mexican economy, this sector is not without difficulties.[32] PEMEX, owned by the government, is the world's third largest corporate producer of petroleum and also the most highly taxed and indebted oil company on the planet. Net losses for the company amounted to approximately $3.8 billion in 2005.[33] Despite this, revenue from oil sales accounted for about one-third of the Mexican government's budget in that year. As a share of the country's GDP, oil income accounted for 6.7 percent in 2001 and rose to 8.7 percent in 2005.[34] PEMEX is more important for its role as the government's cash cow than for the number of workers hired by the company, which represents only about one-third of the total number of U.S.-bound migrants in a single year—about 130,000 people. PEMEX sales amounted to about $97 billion in 2006, with $79 billion of that going to government coffers.

The Cantarell oil field in the Gulf of Mexico has been responsible for producing about 60 percent of the country's oil output, but its production level has fallen from 3.4 million barrels a day in 2004 to 3.26 in 2006. Further, PEMEX officials revealed in 2005 that they had overestimated oil reserves in the region by as much as 53 percent.[35] They announced in 2006 that the Cantarell field, the world's largest producer after the Ghawar oil field in Saudi Arabia, is threatened by encroaching water and gas, which contaminate oil and make it more difficult to sell. Production is predicted to decline 6 percent by 2008. Not only could this represent a threat to Mexican financial stability, but it could also crimp U.S. inventories and contribute to a spike in oil prices internationally. Canada and Mexico are the States' two leading suppliers of oil.[36]

Given the United States' historical interest in Mexican oil, which was accentuated especially after the 2003 invasion of Iraq, one would naturally expect Washington to assume an active interest in security for Mexico's petroleum sector. The shadowy outlines of a key U.S.-Mexican security relationship regarding oil surfaced in the public realm in November 2004. At that time, it was abruptly announced by Mexico's secretary of energy, Fernando Elizondo, that the U.S. military had taken over surveillance and protection of Mexican oil installations in the Gulf of Mexico. Caught off guard, PEMEX functionaries lamely asserted on 8 November 2004 that the company would not officially recognize U.S. military vigilance over its operations in the Gulf, because that important agreement was made without consulting the Mexican public and the Mexican Congress. Consultation or not, U.S. military protection of the installations continued.[37]

Since then, the Mexican military slowly has become more directly involved in the security of the petroleum industry. Conrado Aparicio Blanco, the director of the country's Center for Superior Naval Studies,[38] suggested that by early 2006 there were three primary security foci for the Mexican navy: the protection of petroleum installations in the Gulf of Mexico, narcotrafficking, and the effects of extreme weather associated with global warming. "We have focused much of the resources we have clearly on petroleum resources," he indicated, adding that the navy found itself short of the resources necessary to do the job properly. He noted that although the navy had purchased Israeli missile launchers and surveillance equipment in 2004 and 2005, its attempt in 2005 to purchase a fleet of Russian aero-naval Sukio-27 and Grippens failed due to budgetary constraints.[39] Any gap left by the Mexicans regarding petroleum security are likely to be filled by the Americans.

By way of comparison with Colombia, it is worth emphasizing that Mexican oil production has not been socially contested to the point of strategic crisis. In Colombia, it will be recalled that the petroleum workers' union has been the target of severe repression due to its stalwart position rejecting privatization. Privatization of PEMEX has remained taboo in Mexico. Rather than exacerbating strategic woes, so far the revenue generated from PEMEX indirectly may have alleviated some security concerns. Oil wealth has helped reduce Mexico's debt loads and has also provided the country with substantial currency reserves that can insulate it to some extent from potential financial shocks. For Mexico, the central worries are how much longer its petroleum reserves will hold out, and whether or not it can provide a level of security in this sector that does not provoke incursions against its sovereignty through U.S. intervention.

The RMA, Energy, and New Economic Spaces: La Parota and Plan Puebla Panama

There are other themes related to Mexican energy production that are highly problematic in the strategic realm. With respect to social struggle associated with the energy sector, perhaps "Exhibit A" is the La Parota hydroelectric project in Guerrero, which is destined for completion in 2010. Located about 30 miles from the Acapulco airport in Guerrero, it entails the construction of a 532-foot dam that will produce 900 megawatts of electricity. The facility is planned to supply electricity to Acapulco, and to connect to the Electricity Interconnection System for Central America, which involves a 1,830-kilometer energy transmission line between southern Mexico and Panama.

Financed largely by the Inter-American Development Bank, and involving a mix of state and private capital, the electrical transmission line is designed to supply power to new private development projects in southern Mexico and Central America. As a result of the project, the cost of electricity is predicted to drop from about 11 cents per kilowatt-hour in 2004 to 9 cents in 2010. It was the most advanced project of PPP in 2006. The plan, which will be discussed in greater detail below, represents an emerging trade and investment scheme implemented in 2001 to develop the region by using cheaper labor than available in central and northern Mexico, thus rendering the area better able to compete with China.[40]

Yet the PPP, and especially La Parota, have outraged many in southern Mexico. Official estimates suggest that the dam will mean the flooding of 170,300 hectares of land, thus affecting 590 homes, with a total displacement of 2,812 people. In sharp contrast, local protestors claim that up to 20,000 to 25,000 people actually will be displaced along the Papagayo River watershed.[41] Although a protest involving 3,000 people against the general outlines of La Parota occurred in March 2003, demonstrations since then have been more focused.

Much of the social struggle has centered around land expropriation from *ejiditarios,* that is, farmers on communally owned land. Since the controversial 1992 alterations to Article 27 of the Mexican constitution, *ejiditarios* have been permitted to sell their land privately. Only a simple majority of signatures from *ejiditarios* is required to permit the expropriation of their community's land for the dam's creation. But sizeable minorities have been extremely disgruntled, as was the case in 2005 in Dos Arroyos, where 343 of 572 farmers agreed to sell. Other disputes centered on the expropriation of land where allegedly just a minority approved of the sale, such as in Cacahuatepec, where apparently only 823 of 2,300 signatures were obtained.[42] Many were injured in both of the protests noted above. Importantly, the leader of a prominent group opposed to La Parota was assassinated in January 2006. This is in addition to four other similar deaths in 2005. Further examples of repression include the infliction of severe injuries to three protestors and four illegal detentions in that year.[43]

Thus, the dispute over La Parota has pitted community members against one another and has been highlighted by violence. Besides the debate regarding the sale of community property, another contentious issue concerns job prospects associated with the project. Supporters of La Parota were lured not only by the monetary value of land sales, but also by the promised creation of 10,000 jobs involved in the dam's construction.[44] This is highly significant, given the pronounced lack of solid employment opportunities in Guerrero. Potential job opportunities have led to the endorsement of the project by some important labor unions, for example, the Sindicato Unico de Trabajadores Eléctricos de la República Mexicana, which represents electrical workers in the country.[45]

Other points of angst centered around La Parota's planned provision of electricity to Acapulco and to the Central American power grid, but not to many interior regions of Guerrero. The EZLN's Marcos called La Parota a development plan for "the tourist jet set," rather than for the masses of impoverished peasants in the state.[46] The indigenous of Guerrero have the least access to electricity in the country, and fare even worse than their compatriots in Chiapas, as demonstrated by Table 7.1.[47] Given their concentration in the state's interior region of La Montaña, they will not likely benefit from energy generated at La Parota. Strikingly, neither will those peasants who will be displaced by the dam.

Subcomandante Marcos, along with a delegation of the Zapatista's national road show called the "Other Campaign," framed La Parota as a centerpiece for social struggle during his lively and provocative tour of Guerrero in April 2006. He told thousands of Guerrero locals, "We all know well what the dam will mean for the land—destruction and death." Deeming the project to be a "crime," and suggesting that the EZLN and others jointly fight any military imposition by the government to enable construction of the dam against popular wishes, Marcos succeeded in raising public awareness of the issue both globally and in Mexico.[48]

La Parota was perhaps the most visible component of Plan Puebla Panama by 2006. It is noteworthy that the PPP itself has sparked widespread social rebuke since 2003. Let us explore why this development project has generated so much public concern and has evoked the specter of security threats. The PPP was conceived in the late 1990s, in the context of the Zapatista uprising and the political awakening of southern Mexico by critical NGOs, indigenous groups, and others. President Zedillo had ordered Santiago Levy, Sub-Secretaria de Hacienda y Crédito Público (Sub-Secretary of Housing and Public Credit), to concoct a development plan for the impoverished southern portion of the country. This represented a top-down, "now hear this" approach to development. The resultant PPP was initiated as a proposal in September 2000, and then implemented in March 2001 with the signing of an agreement by President Fox and the Central American presidents.[49] The development plan should be viewed in the context of what would emerge as the Central American Free Trade Agreement (Cafta). It embraced the wider neoliberal components of privatization, deregulation, and resource extraction largely for Northern consumption. It also entailed the construction of maquiladoras offering lower wages than those straddling the U.S. border, which averaged around $2.50/hour, to compete with China's typical 45 cents/hour.[50] About three-quarters of Mexico's indigenous population live in southern Mexico, and are the most likely to witness PPP's acute effects.

As outlined by the Inter-American Development Bank, which has largely financed the project, the PPP entails eight initiatives: sustainable development, human development, prevention of natural disasters, tourism, the facilitation of commercial interchange, integration of roads and other transportation, electrical interconnections, and bolstered telecommunications.[51] More telling is the distribution of funding among these projects, with highway construction allotted 85.2 percent of the total budget, electrical connections 11.1 percent, and the remaining 4 percent split among the rest. Sustainable development received just 0.4 percent of the total budget, for example, and human development only 0.8 percent.[52] In essence, PPP is about highways and electrical grids, which are designed to provide basic infrastructure for private industry in the sectors of resource extraction, tourism, and manufacturing. Like Nafta and Cafta, the plan draws Mexico into regional forms of globalization. Illustrative of this is the refashioning of Mexican roadways to reflect regional rather than national production. That is, the PPP will orient Mexican highways away from their current system based on modern national centralization, where all paths seem to lead to Mexico City. By contrast, PPP's 3,159 kilometers of highways, of which 1,007 are in Mexico, will run north to south to facilitate regional trade throughout Mexico and Central America.

Almost as soon as it was launched, the project began to wilt. Six months after its initiation, the September 11 attacks took Washington's focus off the PPP and away from Latin America in general. A mild recession in 2001–2002, precipitated by a contracting maquiladora sector and by a chronic agricultural crisis, reduced

Mexico's budget for the plan. At the same time, the Inter-American Development Bank was unwilling to provide Mexico with preferential loan rates, available to poorer Central American countries. Further, the Mexican private sector has not stepped up to the plate regarding anticipated investment funds, suggesting that they harbored doubts regarding the profitability of the project. In relation to this, they may also have been alarmed by the loud wave of protests apparent since the PPP's announcement, in addition to potential security threats given the presence of guerrilla activity in the region.

A marathon of public demonstrations against the plan commenced in 2002 and continued into 2003, but has dissipated since that time. What was the fuss about? NGOs in southern Mexico have feared that the PPP is a neoliberal development plan that may benefit TNCs but may be detrimental to the interests of locals. This should be viewed within the context of negative repercussions that many in southern Mexico associate with Nafta and globalization. Among these are the country's worsening agricultural crisis, a local perception of race-to-the-bottom wages, and ill effects linked to the privatization of communal or state property.[53] Further, with good reason, many such NGOs in Oaxaca and Guerrero claim that the process was not a consultative one. Instead, it was perceived to represent an authoritarian imposition of a development model meant to benefit "TNCs, oligarchs and international financial institutions."[54] The local poor were cast as the "objects of development, never its subjects."[55] On 12 October 2002, Indigenous Peoples' Day, 60,000 Mexican Indians blocked roads and led other forms of protest against the PPP. Over 400 indigenous women from the Coordinadora Diocesana de Mujeres demonstrated against the plan in May 2003.

Opposition to the project has continued since that time, such as pronouncements in 2006 by the Zapatistas' "Other Campaign," whereby Marcos proclaimed that "the isthmus is not for sale."[56] Generally, protests against the PPP have withered since 2003, due to signals of ambiguous and waning commitment to the project on the part of the government. For example, Mexico slashed its budget for the PPP by $491 million in 2003.[57] No doubt the less than robust economic atmosphere in the country contributed to the government's decision, in addition to other factors mentioned above. But substantial social protests presumably represented another important consideration. It is most likely that the Mexican government will press on with a scaled-down or decelerated version of the PPP, but will do so in a less publicized manner, in an attempt to evoke less public mobilization against the project.

* * *

Part of the RMA concerns the creation of new economic spaces that spell fresh strategic concerns. So, too, do related shifts toward privatization and the introduction of energy megaprojects to meet transnational interests. Certainly, the La Parota project is steeped in the doctrine of privatization. It entails the private sale of communal land, funding from private sources, and the provision of electricity for private interests in the realms of resource extraction, tourism, and manufacturing. All this has provoked an angry outcry among elements of Oaxaca's and

Guerrero's civil society, who fear the plan will benefit moneyed private interests at the expense of the majority population. There is nothing inherently wrong with development that involves private capital, and, indeed, there is much to be gained. But given the well-founded perception that the post-1982 wave of privatization has not necessarily benefited the country's majority population, projects involving privatization are now under stronger pressure to demonstrate their clear value to the communities in which they operate.

Given that it has been more than a generation since the emergence of neoliberalism in Mexico, social forces are attuned to its dangers and have organized politically in a process that demonstrates their increasing maturation. While much of this protest will likely continue to be peaceful, it is also important to recognize the presence of Mexican guerrilla activity in the geographic region of PPP. The EZLN was able to create considerable turbulence during the first couple of years of Nafta, and empowered itself through the attention it received. This suggests that new regional trade arrangements such as the PPP can represent tempting and rather easy targets for insurgents such as the Zapatistas and the EPR. Further, in Colombia, Nigeria, Iraq, and elsewhere, it has become clear that energy megaprojects can represent valuable strategic targets for subversive groups. This may raise concerns regarding La Parota. And given the assassination of leading organizers against the dam—an approach sadly typical in Colombia—there exists the possibility that violence could quickly spiral upward.

Finally, La Parota and the PPP, along with the tourist industry and the petroleum boom, have been presented by the government as bright spots in the country's economy. Although there has been a steady barrage of reasoned public criticism of the projects, it is too early to say with certainty whether or not the PPP and La Parota will provide significant benefits for the majority population of southern Mexico. What is clear is that both these schemes have been implemented in an authoritarian manner that defies participatory models of development. "Bright spots" in the Mexican economy have been insufficient to absorb Mexico's growing army of surplus labor. To that important extent, Mexico's model of development since 1982 has failed. This is the context for mass migration, the swelling illicit economy, the proliferation of insurgent activity, and an assortment of strategic problems associated with them.

Social Forces and the RMA: Peaceful Struggles for Human Security

So far, the discussion of Mexico in relation to the RMA has focused on guerrilla warfare, migration, surveillance, crime, and an assortment of strategic effects associated with energy production and privatization. Now we shall shift to a focus on the empowerment of social forces through the establishment of a strong political voice vis-à-vis human security. Much has been written on the general topic of the emergence of civil society, especially since the 1985 earthquake in Mexico City demonstrated that social forces could work wonders if they asserted themselves and defied expressed government demands for complacency.[58] There has also been much research on the development of civil society associated with the strong political voice established since 1994 by the EZLN and its network of supporters. In some ways, the widespread international attention garnered by the Zapatistas made the project of constructing a civil society with a political voice look easy. By contrast, the underresearched cases of Oaxaca and Guerrero demonstrate the considerable obstacles that are faced in this regard by a marginalized and fragmented civil society.

From Fragmentation to Unity: The Coalescence of Oaxacan Civil Society and the Struggle for Security

The discussion above noted the concerted struggle in parts of southern Mexico against the PPP. This endeavor served to unite the notoriously fragmented civil society in Oaxaca. In a state long dominated by the PRI and exaggerated cronyism, the clearest manifestation of civil society had been an assortment of scattered and sometimes feuding NGOs. Leaders and members of these NGOs have openly acknowledged the deep-seated divisions among them, as well as the paradox that they have shared similar views regarding the nature of the problems they have faced. Among these are the oppressive *cacique* system and related political exclusion. Also important in this regard was their common plight during widespread military repression in the late 1990s, with flashes of this continuing into the new century. That occurred during keen government efforts to neutralize the EPR and its offshoots. Another important part of the context was the climate of economic marginalization, much of this as a result of an agricultural crisis twinned with neoliberal cutbacks in state support. But at least up to 2003, many NGOs, even those concentrated near the city of Oaxaca, cited intense fragmentation among them concerning territorial influence and competition for funding. Particularly divisive were debates concerning forms of struggle. Some groups, for example, favored roadblocks and the physical occupation of prominent public spaces, whereas others preferred what they considered to be less intrusive forms of protest.[59] These political divisions were exacerbated by a tendency toward disorganization among many NGOs.

A point of inflection occurred in 2003 with coordinated protests in Oaxaca against the PPP. Issues surrounding the plan galvanized indigenous groups and a wide array of NGOs throughout the state. Questions concerning land rights, resource extraction, energy plants, major transportation corridors, and other issues associated with the PPP sparked a rallying cry for elements of Oaxacan civil society to join together in resistance. While these protests were centered in Oaxaca, with the epicenter of organization being the Tehuantepec's Centro de Derechos Humanos Tepeyac in southern Oaxaca, groups in Guerrero and Chiapas also banded together with their Oaxacan counterparts. The PPP was so multifaceted that it struck cords among many diverse elements of civil society, and therefore served to unite social forces that historically had been quite fragmented. While protests against the PPP faded in 2004 and beyond, Oaxacan civil society enjoyed the successes of unity as well as of placing the federal government on the defensive. This was instrumental in establishing the basis for the widespread protests in Oaxaca that congealed with the initial protests of teachers in the spring of 2006.

The Oaxaca teachers' union, with about 70,000 members, initiated what had been their almost annual strike on 22 May 2006. As in previous strikes, they occupied the Zócalo, or central square, of the city of Oaxaca—one of the country's tourist magnets. This appeared to be the usual ritual, which was typically settled by a yearly pay raise that meant the picketers vacated the Zócalo in time for the arrival of tourists in late June. But that year's strike turned out to be different. Ulises Ruiz, the PRI governor of Oaxaca, who narrowly beat his leftist PRD opponent in a 2004 election, aimed to teach the teachers a lesson, as it were. He ordered state forces to violently remove the protesters from the Zócalo on 14 June 2006—on the eve of what would have been the height of the tourist season. The teachers, already fuming from a lack of progress in negotiations with the state and from their escalating war of words with the governor, energetically resisted the state's use of force. The result was a melee in which four people died and 92 were injured. This was an unmistakable point of inflection, which transformed the localized teachers' strike into a more general protest among Oaxacan civil society against Governor Ruiz. This wider movement called itself the Asamblea Popular de los Pueblos de Oaxaca (APPO), the Popular Assembly of the People of Oaxaca. It represented the coalescence of the state's civil society to a degree not seen for decades.

APPO demanded the resignation of Governor Ruiz, and its loose-knit but fervent mixture of marginalized members of civil society made numerous demands associated with improved social welfare and the dismantling of neoliberal structures. These demands were as varied as the members of the 300 or so organizations that composed APPO, and ranged from specific requests made by teachers to the broader insistence on the complete dismantling of capitalist structures made by groups of socialists. Perhaps the chief commonality among APPO members was the anger generated by their status as the dispossessed. They resented the strong-armed tactics and historical domination of the PRI in Oaxaca, and were troubled by what they viewed as the illegitimate election of President Calderón in the summer of 2006. Many suffered from the economic consequences of an agricultural crisis that had plagued Mexico's largely indigenous south, and felt victimized by the increasingly expensive and dangerous prospects of migrating to the United States in a desperate search for work. One could go on with a roster of well-substantiated grievances.

Rather than wilting, as the government had hoped, APPO seemed to strengthen throughout the summer of 2006 and into the fall of that year. One obvious consequence was the collapse of the tourist sector in the highly popular city of Oaxaca. Hotels there lost an estimated $150 million by August, and had laid off at least 1,000 employees by that time. Besides the occupation of the Zócalo, considerable portions of the city's colonial architecture featured pro-APPO graffiti. Key highways near the city of Oaxaca were blockaded at various times. In flashes of violence on the part of "loose canons" that supported APPO, at least 12 buses were burned in addition to an assortment of similar incidents.

While much has been made of the significance of the Internet in relation to social struggle, especially with regard to the EZLN, here it is crucial to emphasize the importance of the radio in the Oaxacan struggle. While the wealthy and some NGOs may be "wired," the popular classes of southern Mexico typically do not have home-based Internet access and would have to go to rather expensive Internet cafes.

APPO members and supporters were mobilized chiefly by the radio—a means of communication accessible to the majority population of the poor. By the summer of 2006, APPO had taken over five radio stations, which seemed to precipitate snowballing membership. Here one may draw parallels to Hugo Chávez' Bolivarian Circles—neighborhood support groups that have often communicated through tiny and local radio stations capable of transmitting only to small communities.

A key turning point occurred on 29 October 2006 when federal police officers entered the city of Oaxaca. About 4,000 officers forcefully took control of the Zócalo and expelled APPO supporters from the town square. Shortly afterward, about 140 APPO leaders and supporters, including the movement's top brass, were arrested and placed in high-security prisons.[60] Seizing on what it calculated was a good political opportunity, an offshoot of the EPR guerrilla group, El Movimiento Revolucionario Lucio Cabañas Barientos, claimed responsibility for the placement and detonation of three bombs in Mexico City in early November 2006. These were placed at sites including the Office of the Tribunal Electoral del Poder Judicial de la Federación, which had presided over the vote count leading to Calderón's victory, and the resort town of Ixtapa in the state of Guerrero, where President Fox and president-elect Calderón were expected to meet. It is not clear how much popular support the guerrillas had generated from the Oaxacan struggle. It is likely that this bombing episode represented failed political opportunism on the part of the rebels, as they had not demonstrated any considerable popular support up to that event or in its wake. Nevertheless, this episode underscored the continued presence of guerrilla forces in the south, and placed a spotlight on the increasingly explosive situation in southern Mexico on the eve of President Calderón's inauguration. By early 2007, the Oaxaca teachers' struggle had left 20 dead, 381 hurt, and the continued detention of 56 of the original 336 detainees. A nightmare situation for the new president was a political conflagration in the south that united organized resistance movements that had existed in Chiapas since 1994 with newly organized and explosive manifestations of struggles in Oaxaca and Guerrero. Let us turn to the case of Guerrero, then, and discuss the evolution of organized resistance movements in relation to the RMA.

The Emergence of Guerrero's Civil Society and the Struggle for Political Voice

While a rough patchwork of civil society in Guerrero was detectable in the 1980s and early 1990s, a decisive turning point appeared with the formation in May 1994 of the OCSS. It sprouted with about 600 members in an agricultural region just north of Acapulco, the area where Cabañas had originally launched his struggle. OCSS members were obviously inspired by the watershed emergence of Chiapas' EZLN in January of 1994. The massacre of 17 innocent OCSS members at the hands of the Guerrero police in June 1995 led to a crescendo of public outrage that, in turn, sparked development for civil society.

In the immediate aftermath of the massacre near the community of Aquas Blancas, there was a deliberate cover-up campaign led by Guerrero's governor

Rubén Figueroa. The government initially provided to the media a blurry video that appeared to suggest that a peasant member of the OCSS initially fired at the police from the truck in which his group was riding. The truth behind the massacre was revealed almost eight months later when a clear and undoctored video of the melee was shown on television. It depicted the police ordering the 17 OCSS members off the truck, systematically shooting them all until they were dead, and then planting weapons into the hands of the corpses to make it look as if they were the aggressors and the police were acting in self-defense. Governor Figueroa was later held responsible for the massacre by Mexico's Supreme Court. In 1997, 28 policemen who had assassinated the *campesinos* were sentenced to 18 years in prison, along with two high-ranking officials with the Figueroa government who received sentences of 19 years.

A founding member of the OCSS, Hilario Mesino Acosta, explained that the group was formed to confront a number of issues facing the majority of peasants in the region. Leading the list was mounting economic difficulties, especially as a result of an agricultural crisis related to Nafta provisions. He also cited extreme political exclusion under an authoritarian state government, political bullying by local *caciques* and their *pistoleros* or hit men, and ill effects associated with illegal logging.[61] Remarkably, these were the same elements that underpinned Cabañas' struggle discussed in Chapter 5.

Mesino Acosta also pointed to a series of human rights abuses committed by authorities who obviously suspected that the OCSS was simply a political "front" for the EPR guerrillas. That is, given the historical presence of guerrillas in the state since the 1960s, and the debacle created by the emergence of the Zapatistas, the last thing the government wanted was a resurgence of guerrilla activity in Guerrero. Mesino Acosta, and other OCSS members who joined him in an interview, observed that an intense militarization of Guerrero had commenced since the appearance of the EPR in 1996. Much of this, they said, was conducted under the guise of fighting narcotrafficking, with the military concentrated in areas where opium poppy cultivation was known to take place. Mesino Acosta and other members of the group insisted that there had been no connection between their group, which peaked at 800 members, and guerrilla organizations.[62] The peaceful protests of the OCSS have been manifested, for example, through public demonstrations and through roadblocks aimed at illegal wood harvesting. The group has attempted to work in concert with other NGOs in Guerrero, but with mixed success—a point to which we shall return.

Since the formation of the group in 1994, Mesino Acosta identified two major contextual transformations. The first involved the government's post-9/11 use of the discourse on terror to fight subversive forces in Guerrero, even though the actors and conditions there had not changed much since the 1970s. We observed the same trend in Colombia. The second major development concerned increasingly better organization of the disparate NGOs in the state of Guerrero.[63] Although progress had indeed occurred on that front, some obvious limits remained. The OCSS, for example, had no office equipment, computer, or telephone, and was reachable only through the mailing address of some of its members.[64] Further, while the issues raised by the OCSS were highly relevant, in general the organization lacked the capacity to articulate its concerns in a manner that could generate sustained attention both nationally and globally—a point to which we shall return in the "Conclusion."

A major development in terms of the fortification of Guerrero's civil society occurred with the formation in 2001 of La Red Guerrerense de Derechos Humanos, an umbrella organization designed to link five disparate human rights groups in the state.[65] A clear highlight of the group's accomplishments was the arrangement of a formal testimony of various Guerrero human rights organizations before the UN High Commissioner for Human Rights, held in the state capital of Chilpancingo in May 2003. At the same time, it released a formal report documenting Guerrero's legacy of repression since the 1970s. It noted the government massacre at El Charco, which will be discussed later in this chapter, and the harassment of human rights leaders such as local hero Hilda Navarrete. It highlighted what it called "systematic aggression" against civil society, including forced detentions and intimidation, and lamented the well-publicized murder of Digna Ochoa, the celebrated lawyer mentioned earlier who had defended EPR members as well as local ecologists fighting illegal wood harvesting.[66] It is important to emphasize that this was the first time that Guerrero's nascent civil society expressed a collective voice to a major global organization and, in so doing, successfully demonstrated that the problems that exist in the state are as serious as those in Chiapas.

Guerrero's NGOs are diverse not only in terms of region but also in terms of their organization. Some NGOs, such as the Comisión de Derechos Humanos La Voz de los Sin Voz, are part-time affairs. This organization has been directed by Hilda Navarrete, who simultaneously has worked more than 40 hours weekly running a small and inexpensive restaurant. A true maverick, her human rights work has spoken on behalf of those who populate the region where Cabañas' struggle began. Despite her constant and tireless efforts, which are revered in the state, her organization, and others like it, naturally is limited by a lack of full-time staff.

A similar group in the same area is the Centro de Derechos Humanos Mahatma Gandhi. Its director, Francisco Díaz González, indicated that the region traditionally has grown coffee as one of its major legitimate crops, but has been suffering due to years of near record-low coffee prices. Besides economic woes that have catalyzed local support for guerrillas since the time of Cabañas, government repression escalated between 1996 and 2004 in an attempt to eradicate subversive forces linked to the EPR. The director said that any locals wearing boots and headed toward the mountains were immediately suspected of guerrilla activity. He also indicated that a wave of spies had infiltrated the community and that a campaign of intense militarization of the region was designed to frighten the population from organizing politically. He noted that, despite these impediments, the most significant positive development he has witnessed since the early 1990s has been the emergence of a "web of human rights agencies" in Guerrero since the formation of the *Red* or "network" in 2001.[67]

By contrast to these valiant but smaller agencies, the state's most potent, best organized, and well-staffed human rights organization is located exactly where one might not expect to find it—deep in the mountains of the La Montaña region.

It entails a six-hour bus journey from Acapulco, or a four-hour excursion from the state's inland capital of Chilpancingo. Illiteracy in La Montaña is measured at 71 percent.[68] This is the heart of the state's indigenous population, and includes the Me'Phaa, Naua, Na Savi, and Amuzgo communities. It is also an area that has relied heavily on corn cultivation, and has been under siege by agricultural provisions associated with Nafta. Emigration from the area, largely as a result of the local agricultural crisis, increased 40 percent between 2001 and August 2005.[69] Further, the sometimes volatile La Montaña region has hosted intense guerrilla activity as well as narcotrafficking, and there have been frequent flashes of military repression there. Through the efforts of its dynamic director, anthropologist Abel Barrera, the Centro de Derechos Humanos Tlachinollan (CDHT) was established in 1994 in Tlapa de Comonfort. The agency has had a professional and full-time staff that includes lawyers, international communications specialists, publicists, graphic artists, and others. Many staff members have been volunteers, or have received funding from other organizations such as churches, allowing the CDHT to operate effectively on a limited budget.

The central task of the CDHT has been to awaken global interest in Guerrero's human rights drama against the backdrop of local political exclusion and the *cacique* system. Its director, Abel Barrera Hernández, suggested that the organization's job has been to overcome the state's current status as "forgotten"—a frustrating predicament given the barrage of global attention devoted to Chiapas. The agency also has been challenged by the ramifications of Guerrero's relatively violent political culture, where individual and group feuds often tend to be settled by violence, where animals are sometimes sacrificed at fiestas, and where the government routinely has used violence as an instrument to protect authoritarian and exclusionary rule.[70]

Despite a difficult environment, the CDHT has earned a long list of kudos. It has been successful, for example, at minimizing the violence emanating from land conflicts between indigenous groups, and at representing the interests of Policías Comunitarias, which will be discussed shortly. It also has succeeded at shining the floodlights on examples of state repression in efforts to reduce their occurrence. For example, it has hosted members of the UN Commission on Human Rights in addition to an assortment of other high-placed global agencies. The key element of the CDHT's success has been its ability to articulate the plethora of problems facing Guerrero in a manner that grabs the attention of national and global figures. It has worked hard to place itself on the cutting edge of the RMA in terms of the themes it addresses, such as security issues associated with the new economic space of the PPP, ecological concerns and an appreciation of the biosphere, identity politics, and the ill effects associated with narcotrafficking.[71] Tlachinollan has acknowledged the advent of the RMA by confronting these themes through networked organization, a reliance on global communication systems, and the articulation of issues in a manner that attracts the global gaze.

Overall, the CDHT and other NGOs represent the dawn of what could become a strong and activist civil society in Guerrero. Although by 2007 they still represented only a patchwork of disparate and sometimes divided NGOs, they have begun to give voice to the majority population repressed by an exclusive and authoritarian government. They are just beginning to discover the power of networked organization, a central component of the RMA and a feature plainly apparent in Chiapas. Ironically, it was the murder of 17 innocent members of the OCSS by the government that helped to provide a thrust toward the development of such a network. We shall return to a consideration of these NGOs at the end of the chapter when we examine what is required to attract the power of the global gaze. Let us now turn to an analysis of other social resistance movements that have attempted to assert power through rather bold though peaceful reclamations of political space.

Identity and Biopolitics

Power passes through individuals. It is not applied to them.[72]

Michel Foucault, Society Must Be Defended

Identity politics entails the regrouping of political loyalty away from the nation-state, as expressed through the rubric of patriotism that was dominant during modernity, toward new forces of political camaraderie with distinct spatial arrangements. With regard to Mexico, identity politics has been on the ascendant since at least the 1980s, and has been manifested through gendered politics, ecological movements, religious movements such as those associated with liberation theology, and so on. Especially since the emergence of the EZLN in Chiapas in 1994, the indigenous have been an important element of the country's identity politics. A crucial watershed for that movement was the 1996 failure of the San Andres accords, which would have provided various forms of autonomy to the country's indigenous population. But this setback spawned other highly positive advancements. Overall, as social forces seek to define new identities against the backdrop of globalization, we shall see that they harken back to a consideration of Mexico's origins that privileged the indigenous and *campesinos.*

Caracoles

Before returning to the case of Guerrero, it is worthwhile here to take note of a relevant development toward autonomy in Chiapas. Frustrated by the total lack of progress with regard to government negotiations and policies, the Zapatistas took matters into their own hands with the creation in 2003 of *Caracoles* (the Spanish word for "snails"), or "autonomous rebel Zapatista communities." Reflecting on the establishment of five Caracoles, its members emphasized that "we have not sold out or surrendered, we have made progress."[73] This represented a fresh manifestation of what Foucault deemed to be "biopolitics." During modernity, this concerned how a social body was to be governed within the framework of a national territory. It involved the endeavor "to rationalize the problems presented to governmental practice by the phenomena characteristic of a group of living beings constituted as a population: health, sanitation, birthrate, longevity, race."[74] By shifting biopolitics from the national to the local level, the Caracoles established their own health centers, schools, economic cooperatives, and other institutions. They represent truly autonomous government—a refashioned social body governed by the indigenous.

Members of a Caracol in Oventic, Chiapas, which was created in August 2003, indicated that the motivating force behind this autonomous zone has been "the quest for autonomy" and the drive "to live up to the San Andres accords."[75] Their first priority has been health, with the second being education. Doctors and nurses from Mexico and abroad have volunteered their services to the Caracol, which, residents say, provides better health care than would otherwise be available from the government in neighboring communities. Members of the Junta de Buen Gobierno de Oventic said even nonresidents of their Caracol have sought medical services there, where any payment made is by donation. To make ends meet, this particular Caracol created a language school for international students, who pay $144 for a three-week course. Students live on-site and are educated in Zapatista politics. The Caracol also contains an organic corn-and-rice farm, which shuns genetically modified plants. But residents lamented that the quality of their land is low.

Interestingly, the only two questions the Caracol directors would not answer during my interview with them concerned how many people resided there, and, importantly, in what ways life was any better in the Caracol than in other parts of Chiapas.[76] There has been a general exodus from Chiapas since the early 1990s, with an estimated 30,000 to 50,000 Chiapans emigrating illegally to the United States on an annual basis. Perhaps that level of emigration explains in part the sensitivity of the directors regarding the number of Caracol residents. In other words, the creation of these small communities has done nothing to stem the tide of migration. Further, there was no discernable distinction between the standard of living in the Oventic Caracol and that in the surrounding communities—residents of the Caracol obviously lived in poverty. Perhaps the biggest material benefit was the medical treatment the council members claimed was available there as a result of volunteers from elsewhere. The most significant benefit of the Caracol may be political, in that the indigenous residents govern themselves. It is important to emphasize that the existence of the Caracoles is tolerated by the Mexican government, which clearly has the military power to oust them whenever it pleases.

Guerrero's Policías Comunitarias

> We're living among criminals and pygmies, because the big boss only favors midgets who won't stand over him. . . . That isn't what we wanted when we started the Revolution.[77]
>
> *Carlos Fuentes,* The Death of Artemio Cruz

Although they have not received the global publicity associated with the EZLN and its Caracoles, Guerrero's Policías Comunitarias are worthy of attention. They are a community police force entirely separate from the official government forces. PCs began operation on 14 October 1995 and were influenced by the initial success of the Zapatistas. The first PCs were created in Santa Cruz El Rincón in the municipality of Malinaltepec, a largely indigenous community. The PCs were designed to "put into practice" the accords of San Andres. Members of the Coordinadora Regional de Autoridades Comunitarias (CRAC), the directorate of PCs who also serve as its six judges, said they formed "due to impunity, corruption, bribery, and *cacique* politics—there was no real justice, no legitimate security."[78] This suggests that they formed for reasons similar to those that helped launch the Guardias Rojas during the Cárdenas epoch, as noted in Chapter 5. The PCs have provided alternative justice, and have done so in a multilingual, interethnic environment. The leaders stress that their organization represents the link "between economics, equity and security."[79] By 2006, the PCs had established a force of 612 police present in 62 communities. Leaders of the PCs have emphasized that they serve both indigenous and mestizo (ethnically mixed) populations.

The PCs claim to have reduced crime, including assaults, robbery, kidnapping, and sexual assaults, by an average of 90–95 percent in their communities.[80] They carry guns and provide entirely voluntary, unpaid service. The members received limited training from the Comisión Nacional para el Desarrollo de los Pueblos Indigenous.[81] Importantly, they directly challenge the authority of official Mexican police. When I visited the group in 2005 in San Luis Acatlán, their office had a lineup of clients waiting for five to six hours to see these autonomous authorities. The authorities indicated that this is typical. It was also observed that the official governmental police station was empty, which the movement's authorities claimed is testimony to their legitimacy and to the public antipathy toward official police.

CRAC members indicated that they have experienced a "tense" relationship with government authorities, who, they said, viewed them as "illegal and usurping government authority" and routinely attempted to intimidate them.[82] The government has made energetic overtures to absorb the PCs. In 2004, for example, the government of Guerrero attempted to "legalize them by absorbing them into the Municipal Preventative Police and by paying them a monthly salary of about $300."[83] But, sticking to their guns as it were, the PCs resisted what must have been a tempting material offer. On their tenth anniversary, the members accused the government of "criminalizing them" and clearly indicated that they wanted the official police out of their territory.[84] The government was especially critical of the self-styled police after 13 of its prisoners escaped incarceration in June 2005. Guerrero's Procurador de Justicia (attorney general) stated that the PCs were not well defined, were "politically driven," and did not "have a good relation with government in any form."[85]

On their tenth anniversary in October 2005, the PCs released a document reflecting on their plight over the past ten years. The discussion began with a conceptual overview that focused on the "recuperation of territory." It noted that "the concept of territory is that it is a space that is yours." The leaders indicated that they would focus on a "small space," one resistant to national and global forces that privileged "the private over the collective" and that promoted new trade blocs such as the PPP that further usurped local power. They wished to emphasize ecological concerns and saw their space as connected to the biosphere. This stands in sharp contrast, for example, to unpopular megaprojects such as the highly contentious La Parota Dam that is part of the PPP—as discussed in the previous chapter.[86] The agenda of the PCs, then, is far larger than just the establishment of a community police force. It is about creating a new space, with a new "popular government," a new attitude to the biosphere, and, importantly, a new economy that emphasizes an "internal market" that would attempt to protect the economy from the devastating agricultural effects associated with Nafta.[87]

Frustrated by failed attempts to co-opt PCs, the federal government in October 2006 sent military troops into communities within the group's realm. The government justified the action by pointing to a sudden wave of rising crime during September 2006, where the problems centered on increased levels of assault, cattle rustling and other theft. But the PCs had already mobilized to fight the crime wave by placing new checkpoints on key roadways and by assigning more police to specific areas where crime had risen. Once inside the communities, army officials requested the names and addresses of the organization's members, attempted to take their photographs, and so on. This was clearly an attempt to intimidate the PCs and the community members who supported them. The military retracted shortly thereafter due to the aforementioned crisis in Oaxaca that appeared to reach a boiling point during late October and early November 2006.

The PCs are perhaps more successful than the EZLN's Caracoles. Although they harbor broad ambitions for the future, the PCs' initial focus has been strictly limited to implementing and redefining local security. The Caracoles, by contrast, have had a far wider agenda, but have not demonstrated the extent of success in any particular area that can match the achievements of the PCs. Their astonishing growth from an initial presence in a single community to their establishment in 62 localities is testimony to their popularity, as is the steady lineup of citizens wishing to consult with them. While their claims of reducing crime by as much as 95 percent are not verifiable through other sources, and there have indeed been episodes of spiking crime rates during their rule, it is fair to say that they would not have enjoyed the obvious popularity they have achieved without reducing crime rates significantly. Perhaps their most important achievement has been the provision of legitimate authority to a population historically victimized by the corruption of the *cacique* system.

The PCs are part of the RMA from the bottom up, and signal a reclamation of territory by the marginalized. The movement has been based largely on identity politics. It is a classic example of the new biopolitics, where political space is viewed in a novel way, where a new body politic is defined, and where the political rules of security are created to serve those who had been victims of the traditional system. This police force occupies a space between unarmed NGOs and armed insurgents. They have attempted to monopolize the use of force in their communities to provide grassroots security. In so doing, they have usurped the state's traditional role. Although it would be an overstatement to suggest that the Mexican government has been graceful vis-à-vis the PCs, the government has indeed tolerated them. Surely the government has calculated that disarming the movement could create a degree of social tension and violence that could spark a strategic crisis in the region. The PCs are an important social force[88] and have not yet received the national and global attention they deserve.

Grassroots Surveillance: Social Forces and the Global Gaze, a Comparison of Guerrero and Chiapas

We all need witnesses of our lives in order to live them.[89]

Carlos Fuentes, The Death of Artemio Cruz

Let us explore the crucial question of why Chiapas has garnered so much national and international attention, yet the deserving case of Guerrero remains largely in the shadows. As we can see from Table 7.1, the indigenous in Guerrero are doing even more poorly than those in Chiapas with regard to sanitation and access to electricity. They are roughly comparable in terms of education levels, poverty and inequity, although the indigenous in Chiapas fare slightly worse by these measures. However, it is noteworthy that Guerrero ranks last among Mexican states in maternal mortality.[90] Overall, the point is that Chiapas and Guerrero are similar in terms of the marginalization faced by the indigenous populations in these states. But the indigenous in Chiapas, where 27 percent of the state's population speaks indigenous languages, have attracted far more political attention than those in Guerrero, where they represented about 17 percent of the population in 2006. This is reflective of a general pattern whereby the world has been willing to look at Chiapas but has been reluctant even to peek at Guerrero. Why is this so? Surveillance is a vital component of the RMA, as we saw in the last chapter vis-à-vis vigilance at the U.S.-Mexico frontier. In that example, though, the focus was upon the State's use of surveillance. Here we shall stand that on its head, as it were, and examine the use of surveillance on the part of social forces within the context of the RMA. The central question in this regard concerns what it takes for social forces to attract the power of the global gaze.

Table 7.1 A Comparison of the indigenous populations in Chiapas, Guerrero, and Oaxaca

	Chiapas	Guerrero	Oaxaca
Income of indigenous as percentage of nonindigenous	32	36	54
Percentage of indigenous versus nonindigenous in poverty or extreme poverty	68.7 vs. 38.9	63.5 vs. 55.6	55.6 vs. 33.1
Average years of school attendance for indigenous versus nonindigenous	3.0 vs. 5.1	3.1 vs. 5.5	4.0 vs. 5.6
Number of bedrooms per household for indigenous versus nonindigenous	1.5 vs. 1.7	1.4 vs. 1.6	1.5 vs. 1.7
Percentage lacking sanitation for indigenous versus nonindigenous	70 vs. 27	82 vs. 40	65 vs. 41
Percentage with access to electricity indigenous versus nonindigenous	74 vs. 87	61 vs. 89	78 vs. 89

Source: Selected data from the World Bank, Tania Carrasco et al., "Indigenous Peoples in the States of Chiapas, Guerrero and Oaxaca" (World Bank: Washington DC, 2002).

By using multiple forms of media—the Internet, the electronic press, the radio, and so on—it is possible to shine a global floodlight on serious social injustices. This grassroots approach to surveillance can protect resistance movements and guerrilla groups from assaults inflicted by a government that perceives it can operate within the shadows of national privacy. The military is less likely to commit broad human rights abuses when it knows the world is watching. This is especially so when the country is part of a trade bloc such as Nafta, whereby a new spatial dimension of production means that trade partners take a greater interest in each other's political and strategic affairs.

Timing is crucial to surveillance. Messages not only must be disseminated at the right time, but in some cases also need to be dispatched instantly and simultaneously. Real-time use of information is crucial to protection and security and to providing the organizational networking that can boost the power base of resistance forces. Thus, through constant and instant surveillance, the organizational base of resistance movements can increase from small and isolated venues with highly limited recruitment capacity to a huge potential of national and global supporters that provide a useful instrument of power through their collective gaze. Beyond the significance of national support, it is surveillance at the global level that represents one of the best hopes provided by globalization to social forces struggling against military repression, economic inequity, racism, gendered oppression, ecocide, and so on. Further, real-time surveillance can neutralize what Marx and others considered to be the conservative tendencies of peasants in the countryside.[91] This tendency was reinforced by the incapacity of those in rural areas to organize to the extent possible in urban environments, where people were in close contact. Cybernetic surveillance educates and unites those in disparate locations, sometimes with radical results.

* * *

Let us begin by comparing Mexican insurgents, particularly the Guerrero-based EPR and the Chiapas-based EZLN, with respect to their manipulation of the instruments of surveillance. There are some important and relevant contextual distinctions between the two groups. First, the EPR and its offshoots (heretofore referred to only as the EPR for the sake of brevity) did not appear within the context of a major uprising, as had been the case with the ELZN, which mobilized thousands of mostly Mayan Indians. The sudden and substantial uprising of the Zapatistas, no less on Nafta's inauguration day, grabbed media attention in a way that the EPR's protracted, smaller, and less sensational appearances have not.

Next, the EZLN began with a military strategy, but they fought in the open in their attempt to rally a revolutionary march to Mexico City that never happened to materialize. They were contained militarily only 12 days after their appearance. By contrast, the EPR has been a clandestine guerrilla group from the start. In addition to bombing public and private infrastructure, it has battled the military with classic hit-and-run attacks and then retreated quickly into obscurity. That formula has continued in a sporadic fashion through 2007. The apparent reassertion of the EPR's limited military power through the placement of bombs in Mexico City in early November 2006, in an apparent show of support for Oaxaca protesters, presumably has done little to swell the popularity of the guerrillas. The EPR also took credit for the detonation of bombs at two natural gas pipelines in Guanajuato and Querétaro in July 2007, and promised more of the same unless two of its compatriots in Oaxaca were released from what it claimed was clandestine imprisonment by authorities. Like the Mexico City bombs, this too created alarm among Mexican authorities, but bombs that deny ordinary Mexicans access to gas are not likely to generate much popular support for the EPR. The real threat posed by those tactics is the potential to discourage foreign investment due to perceptions of local instability. Overall, the EPR have not been as rapidly contained by government forces as the Zapatistas had been. The EPR did not begin with a blitz that attracted global media, and has operated in a violent and clandestine manner that has been generally unattractive to Mexicans and to the global gaze. Indeed, the most important distinction between these rebel forces is that the EPR is a traditional 1970s-styled guerrilla group that has relied on force, while the EZLN has been a militarily powerless but politically important resistance movement. That is, the Zapatistas are not really guerrillas in the military sense.

Much of the EZLN's success at using surveillance as a tool to exert political power, and to defend itself from obliteration by the Mexican army, can be attributed to Subcomandante Marcos. Although it is important not to diminish the contributions of other members, Marcos' role is outstanding. His charisma, his looks, and especially his intellectual savvy meant he was a magnet for the media and a symbol to mobilize individuals gazing into their computer screens from all over the planet. Marcos has earned a PhD, speaks five languages, can think outside the box, and has a middle-class background—all of which contributed to his prowess at the art of connectivity. The EPR, by contrast, had no such leaders. They were, instead, impoverished *campesinos* who lacked an extensive education and a sense of worldliness. While their intelligence is not in doubt, their limited education and worldview were strikes against them when it came to rallying support outside their local enclaves. They did not possess the wherewithal to craft the cerebrally inspiring and clever messages of Marcos and, rather than political development, pursued a path based on military might.

Further, the EZLN is composed mostly of indigenous members, and this has represented a prominent theme in their thoughtful political platforms. It was one component of their multifaceted analysis that celebrated broadly defined identity politics. Through their novel Other Campaign, which rivaled those of Mexico's traditional parties during a presidential election year, the Zapatistas in 2006 continued to focus on elements such as race, gender, class, age, spatial location, ecology, and others.[92] This stood in stark contrast to the almost monolithic class analysis embraced by the EPR. The policies of the EPR, as we observed in the previous section, were steeped within a framework that was too traditional and "old school" to attract the global gaze.

While the EZLN has presented itself as a united front, the EPR has been fractured to the extreme. It spawned a lengthy and forgettable list of relatively powerless offspring who sporadically planted bombs and ambushed unsuspecting security forces. This represents a serious weakness, because it is impossible to unite prospective supporters when a group is utterly unable to congeal its existing forces. These fractures resulted from feuds regarding the doctrinaire ideology of the EPR, the frustration among its members about zero progress, and the alleged tendencies among some members to engage in crimes such as narcotrafficking and kidnapping. These unflattering elements, among others, did little to equip the EPR with the capacity to attract the national and global support that could have facilitated a disarming political surveillance vis-à-vis the Mexican military.

Overall, the EPR is a *campesino* guerrilla organization dedicated to fighting the inequity and the misery associated with the *cacique* system and with being trapped in Mexico's failing agricultural sector. Haunted by traditional problems that are clear enough, the guerrillas clung to a traditional but unworkable approach. The EPR simply did not have what it takes to attract the global gaze. As a result, the Mexican security apparatus was freer to employ a more intense degree of repression than would otherwise have been the case. For example, an earlier discussion pointed to the government's use of torture to force confessions of links to terrorism. Huge military mobilizations have occurred, complete with mass detentions under illegal conditions. The global gaze failed to register these and other effects of horrendous military repression in Guerrero. Nonsurveyed privacy also has permitted the presence of paramilitary forces in Guerrero to rout out potential supporters of resistance movements. The world turned a blind eye to the social conditions that prompted desperate *campesinos* to take up arms under incompetent leadership in an effort to achieve what was promised by the Mexican constitution almost a century earlier.

For its part, the Mexican government vigorously used surveillance and intelligence to its advantage. Security experts in Mexico indicate that the EPR and its offshoots have been relatively easy to infiltrate, especially through universities in Guerrero and Oaxaca, and to a lesser extent through NGOs.[93] Representatives from most of the NGOs I interviewed in Oaxaca and Guerrero were aware of intelligence operations in their community and with respect to their particular organizations. These included unidentified persons taking pictures at agrarian rallies, strange noises on the phone, which suggested eavesdropping, payments to neighbors of human rights organizations for information on their activities, unidentified persons watching the offices of human rights organizations, and so on.[94]

* * *

This brings us to issues surrounding the ability of civil society to catch the global gaze, especially with respect to social forces in Mexico that have suffered from social and economic injustice. The state of Oaxaca, which like Guerrero had become largely eclipsed by the politics of Chiapas, was thrust into the global public eye in 2006, when an annual teachers' strike ignited into a statewide social resistance movement to protest economic marginalization, social injustice, repression, and *caciqueism*. This point of inflection suggests that the rather sudden appearance of mass movements attracts national and global attention. The trick for social forces in Oaxaca is to sustain the momentum and maintain the attraction of the global gaze. This represents a serious challenge given the mass arrest of social leaders and past tendencies toward fragmentation.

Social resistance groups in Guerrero have faced considerable impediments to appearing on the radar screen for global social networks. The state hosts a civil society that is emerging but that has often lacked a clear voice even on a regional level. Guerrero's indigenous population is not sufficiently large to attract much attention from global NGOs interested in this component of identity politics. It is also burdened by a generally weak education system that does not provide the tools of political articulation.[95] Guerrero's historically entrenched levels of violence, as well as escalating crime such as narcotrafficking, may also have turned off elements of global civil society.[96] Further, it has not attracted the global attention afforded to the sudden appearance of mass movements, as was the case with the EZLN in Chiapas and the APPO in Oaxaca.

There are other aspects of Guerrero's politics that have been dissuasive to global and national support. High on the list is the dispersion of the state's NGOs into regions that can take hours to reach by poor roads, which has resulted in pronounced political fragmentation. This stands in contrast, for example, to the concentration of Chiapas' civil society in the city of San Cristóbal. Further, it was shown that some of Guerrero's NGOs have not had the budget for even basic communication equipment such as computers and fax machines—equipment that is vital both to overcoming spatial dispersion and to disseminating vital information. In addition, civil society there has not been uniformly accustomed to meticulously documenting data related to repression and abuse. This has impeded their capacity to launch an info-war. Moreover, we saw that some of Guerrero's NGOs have been limited by a part-time staff who have full-time day jobs. Finally, infighting and competition among Guerrero's NGOs has weakened civil society in the state, as exemplified in April 2006 when the EZLN's Other Campaign was forced to cancel a scheduled visit to Atoyac de Álvarez due to an embarrassing feud between the OCSS and the Consejo Civico Comunitario Lucio Cabañas.[97] This sad dynamic, as noted earlier, also has been apparent in Oaxaca.[98]

A bright spot here is Guerrero's CDHT in Tlapa de Comonfort. Under the exceptional and excellent leadership of Abel Barrera, its professionalized staff have accomplished the monumental task of attracting flashes of global attention to previously forgotten areas of Mexico, as we observed earlier. Its staff, many of whom are full time, are highly educated and possess the basic equipment and know-how to achieve global connectivity. The group has had a series of superb international coordinators who have encouraged and facilitated global interest in the organization. This represents an important beginning in Guerrero to what promises to be a long and arduous task. However deserving a subject Guerrero may be, in general it does not yet possess many of the features required to attract the gaze of global civil society in a highly competitive arena.

In sum, Guerrero, Oaxaca, and Chiapas represent good comparative case studies to flesh out what does and does not attract the global gaze. The organizational feat of achieving a sudden mass movement will likely attract international media, as demonstrated by the cases of Oaxaca and Chiapas. Further, relatively sophisticated and expensive communication devices, especially computers and Internet, can assist in attracting the global gaze. Technological expertise and related education are required to use these instruments effectively. Thus, there is a clear economic challenge for impoverished social forces to gain access to such equipment and services. Further, a politically united front and a centralized "capital" for information dispersion also help. The messages themselves need to be informative and engaging, and may be augmented by their packaging in forms such as a shiny Web page. There must be an ideological and epistemological sensibility for what is likely to attract the global gaze. Themes such as identity politics, ecology, human rights, and greater social equity have proven to be engaging in key cases. In addition, messages promoting violent struggle are likely to be viewed as repugnant. This has been reinforced by the democratic and peaceful election of the radical Left in various Latin American countries, such as Venezuela, Bolivia, and Argentina. And it does not hurt if a resistance group is led by someone with "star power," such as Marcos. Thus, the effective use of surveillance entails a complex and challenging array of factors.

Conclusion

This chapter analyzed elements of the RMA along two conceptual spectrums. The first concerned a wide span of issues ranging from the broad themes of economic security to human security. The second entailed the tension between the privatization of war, on the one hand, and mounting public resistance movements, on the other. The focus here has been upon an RMA from below in Mexico, especially regarding the attempt by popular forces to resist social and economic injustice through the construction of public resistance movements dedicated to achieving various forms of human security.

The relation between privatization and warfare in both the legal and illicit economic sectors was considered. With regard to the illicit economy, a major strategic worry for both the state and civil society has been the escalating privatized war associated with narcotrafficking. While we focused on the illicit drug trade's place within Guerrero politics, this is truly a national phenomenon whereby related war is fought on city streets, in rural cultivation areas, and on the nation's frontiers. It was observed that narco-warfare in Mexico has included the use of terror. The state has found itself fighting a war against assorted wealthy drug lords in a fashion somewhat reminiscent of Colombia's battle against Pablo Escobar in the late 1980s and early 1990s. Contextually, it is worth emphasizing that the illegal drug industry rose in tandem with the debt crisis and neoliberal restructuring, and has benefited from the relatively porous borders associated with globalized and regionalized production.

It would appear that narco-warriors in Mexico have far more military power than guerrilla insurgents in the country, and are far less reluctant than guerrilla groups to employ grotesquely brutal levels of force. Violence from narcotraffickers has killed nearly 3,500 people in Mexico between January 2005 and March 2007—a number that exceeds the annual deaths associated with warfare in Colombia during recent years. Legalization and regularization remain the only antidotes to the burgeoning industry of illicit drugs, yet the United States has resisted these for assorted political reasons. The result is that the Mexican government will remain on the defensive and will likely continue to lose its battle with narco-warriors. Beyond the political implications of being perceived as fighting a losing battle, the state is highly vulnerable to corruption by narcotraffickers. This debilitates the state even further and has prompted social forces to view the government as increasingly illegitimate. With regard to the RMA, we observe a synergetic mixture of factors—in this case, the blurring of crime, war, and terror against the backdrop of neoliberal restructuring and a government debilitated by corruption.

In the Colombian case, we observed that the trend toward the privatization of formerly state or cooperative property has galvanized security problems, especially in relation to the petroleum industry. A similar story has played out in Mexico. This is evidenced by loud struggles associated with the La Parota Dam project, which relies on the private sale of communal land, and with the PPP, perceived by many locals to represent another neoliberal scheme to advance the private interests of transnational capital. Thus, the push toward privatization has spawned a plethora of security problems because it is perceived by many as not benefiting the majority population. If the state is widely viewed as acting in the narrow interests of private capital, it will likely find itself squaring off against a rising tide of social forces that have grown increasingly hostile to decades of neoliberal failure.

The issue of neoliberal restructuring has been perhaps the most salient catalyst for public resistance forces in Mexico, especially in southern parts of the country—from the ELZN in Chiapas, to APPO in Oaxaca, to the EPR guerrillas and the assortment of counterconsensus groups in Guerrero, and so on. Other related factors that have helped spawn a critical civil society include persistent political exclusion, government repression that has relied on flashes of terror, and corrupt state structures. Social resistance movements in Mexico have been most successful when they avoided violence and therefore did not fall into the government's trap regarding the discourse on terror. This has been the case with critical NGOs, with groups that ave peacefully reclaimed political and strategic space, such as the Caracoles and the PCs, and even with guerrilla movements such as the postmilitary EZLN. Further, critical social forces have met success when they avoided centrifugal tendencies toward fragmentation, when they vigorously explored various dimensions of connectivity, and when they proceeded to form local, national, and global networks. While economic dimensions of globalization have been perceived by many as threatening the majority population, the connectivity and networks available through globalization can empower social forces that possess the skills to navigate this rather competitive new frontier.

Finally, the theme of surveillance in relation to the empowerment of social resistance movements was addressed. With all the fuss associated with the Internet, it is important to underscore that it was the radio, the poor person's communication device, that was instrumental to the organizational success of social resistance forces in Oaxaca in 2006. While the Internet can help link isolated and fragmented groups in disparate geographical regions, as exemplified by the case of Guerrero, it is particularly useful as an instrument to connect local social forces with national and global movements. While the *Panopticon* is a story of how surveillance can serve as an instrument of repression for those in positions of authority, as exemplified on the U.S.-Mexico border, the floodlights of surveillance can also strengthen social forces and protect them against government repression that might otherwise occur if nobody else is looking. While the EZLN almost made the task of harnessing the global gaze look easy, the case of Guerrero drives home how difficult it really is to attract the attention of global social movements in the context of a highly competitive environment.

Thus, various dimensions of the RMA have worked together to explain many aspects of strategic change in Mexico. This chapter has considered the relation between neoliberal restructuring associated with globalization, the privatization of war, the blurring distinction between crime and war, identity politics, the formation of newfangled social resistance movements, and the power of surveillance. Let us proceed to the book's conclusion, where we shall build upon this theme by considering comparatively the cases of Colombia and Mexico.

8

Conclusion

Look not for this year's birds in last year's nest.[1]

Miguel Cervantes, Don Quixote

Despite its typical association with great powers and Middle Eastern politics, the RMA is a global phenomenon with clear manifestations in the Americas. As we observed in Chapter 1, the RMA is very much a work in progress, though its features are becoming clearer. This rupture has entailed asymmetric conflict, complexity, new forms of organization, the use of ultra-surveillance, a variety of themes linked to a globalized political economy, as well as epistemological considerations. This chapter will address how various elements of the RMA have worked together in a synergetic fashion to explain strategic change in Colombia and Mexico.

The Colombian and Mexican cases are interesting prisms through which to view assorted aspects of the RMA, because they bear important similarities and points of distinction. Both host top-tier strategic issues and have been on the front lines of themes associated with the RMA in the Americas. Colombia and Mexico together represent the Latin American epicenter of narcotrafficking and are among the very few countries in the Americas to host established guerrilla movements at the beginning of the new millennium. Burgeoning inequity linked to neoliberalism has spawned or exacerbated security problems in both countries. Further, while PC has entailed the highest level of U.S. military funding in the world outside of the Middle East, on a daily basis Mexico must cope with the intense strategic exigencies of living next door to the United States. Thus, Washington's interests compound the crucial security themes witnessed by both Colombia and Mexico—including insurgency and asymmetric warfare, the link between crime and violence, the privatization of warfare, social struggles related to inequity, and so on.

While Colombia and Mexico share similar strategic issues in a broad thematic sense, these have been manifested differently in the two cases and appear against the backdrop of distinct historical circumstances and contexts. They are distinguished, for example, in terms of geopolitics, political culture, and records of stability, among other themes. And while Colombia has hosted an RMA from "above," one inspired by the U.S. and Colombian governments, Mexico has witnessed an

RMA from "below," one defined largely by resistance movements and other social forces. Related to this, in Colombia, insurgent forces, with their focus on maximizing power through control of territory and resources, often have viewed the population at large to be a nuisance that can be shoved aside through violence, displacement, and other tactics. By contrast, Mexican insurgents and resistance movements have viewed popular support as the essential pillar of political power, even if achieving that support has proven illusive for some groups. Thus, this wide berth of distinctions provides two poles from which to consider various manifestations of the RMA in the Americas.

Continuity, Change, and Episteme

While the RMA signifies a rupture in strategic affairs, there also exist important elements of continuity. For example, issues that have spawned security problems in both Mexico and Colombia, including poverty, inequity, racism, and political exclusion, have deep historic roots. Yet we have seen that the manifestations of social struggles have altered in sometimes-revolutionary ways. This has been a product of the transformation of classic strategic themes that remain as relevant as ever. These include temporal politics, spatial politics, political organization and identity, intelligence/surveillance, good government, and the power of social consent. The paradox is that while there exists a lasting relevance to strategic themes dating back to the writings of Thucydides and Sun Tzu, as we observed in Chapter 1, it is the shifting expression of these themes that is truly revolutionary.

This interplay between continuity and change is exemplified by the notion of asymmetry. This component of the RMA may have seemed rather novel for great powers such as the United States at least until the September 11 attacks, because historically the country has perceived strategic threats to emanate from other major nation-states, as evidenced by the superpower conflict of the cold war. But developing countries such as Mexico and Colombia are no strangers to asymmetry, having witnessed a long trajectory of battles between the state and insurgent forces. What has changed has been the recent manifestation of asymmetry, especially in relation to organization, communication and surveillance, identity and support base, weapons, timeline, and epistemology.

Complexity entails a consideration of multicausality, nonlinear development, and unpredictable consequences for given actions. It is related to ultra-movement during the postmodern era—that is, quicker, more voluminous, and more expansive movement by people, capital, and information. There is more of an emphasis on entropy and even chaos than on the notion of balance that was celebrated during modernity. Strategic actions must be considered in relation to the biosphere. From the specter of sheer political chaos and narco-related ecocide in Colombia, to the pondering of an iron curtain between Mexico and its Nafta trade partner, newfangled complexity is visible in spades. Whether it is the state, insurgents, or civil society, it is ironic that fresh manifestations of complexity are best dealt with through the traditional remedies of adaptability and flexibility. Sun Tzu observed that "water has no constant shape: the ability to gain victory by changing and adapting to the opponent is called genius."[2] Once again, the RMA and its constituent elements such as complexity entail a redefinition of classic strategic themes and associated remedies.

The paradoxical combination of continuity and change is clear with respect to other facets of the RMA examined in this study. For example, we have seen this to be the case with regard to identity politics and related organizational change. Political identity has always been a crucial feature of strategic affairs. Under the RMA there has been an alteration of the homogenous identity of nationalism and patriotism that were linked to the nation-state during modernity. Recently, there has been a celebration of variegated identity politics associated with ethnicity, race, class, gender, and other factors. This is exemplified in the social forces behind Mexico's PCs and also with regard to Colombia's indigenous and black populations in their struggles with displacement and corporate encroachment. Nationalism has not disappeared but rather has been compounded by other forms of identity, as borders change, as the nature of combat transforms, as economic arrangements shift, and as newfangled communication devices link previously disconnected communities.

Regarding other elements of the RMA, it was demonstrated that the political instruments of terror and fear are as old as recorded history. Among other things, what is new in the current RMA is the rather complex "discourse on terror" linked to the September 11 attacks and the subsequent redefinition of global strategic affairs. Further, we have seen that terror has become a favorite tactic among certain insurgent groups and especially among criminal syndicates within the rubric of asymmetric warfare. This has been the case with regard to horrific narco-terror in both Colombia and Mexico and with regard to the terror unleashed by Colombian insurgents. Flashes of state terror in Colombia signal the absence of hegemony, and in Mexico is symptomatic of fading state hegemony. Fear and terror are not new, but some of their recent manifestations have been quite novel.

In a similar vein, surveillance and intelligence have been mainstays of strategic affairs since the writings of Sun Tzu and his predecessors.[3] Profound conceptual alterations of these themes have occurred throughout history, as demonstrated by Bentham's Panopticon. Recent technological developments such as satellite surveillance and Internet connectivity have promoted an atmosphere of ultra-surveillance linked to the current RMA. These developments have spawned implications for a wide variety of players.

Turning to other realms, while political economy always has been linked to strategic affairs, as Thucydides emphasized with regard to Athens' debt crisis and other assorted themes in the Peloponnesian War, the emergence of global transnational capitalism has ushered in fresh security concerns. What is new is that the current RMA is situated against the backdrop of economic globalization. This has spawned unfolding implications with regard to changing borders, altered state structures, transformed political identities, and a quicker tempo for the voluminous movement of people and capital.

An overarching epistemological rupture is the most important element associated with the present RMA. Classic strategic themes are being known in new ways. Throughout the book, for example, it was shown how strategic time and space have been redefined. Traditional political forces such as the state, the church, and the elite media no longer hold a monopoly on the manufacturing of social truth, which is contested by social movements in the realms of spirituality, ethnicity, gender, biopolitics, and so on. This trend is reinforced by the dispersion of information in such a way that it does not have to pass through the filters of the traditional institutions noted above. Weapons of mass destruction and others of the more ordinary variety also have been dispersed, an element that has accentuated organizational shifts linked to asymmetric warfare. All of these components and the synergy between them reflect a systemic rupture—political and strategic "truths" are being conceptualized in fresh ways by various social forces and their new "experts." This is the crux of the "system of systems" approach discussed in Chapter 1. Overall, while thematic elements of strategy demonstrate considerable traces of continuity, the expression of these themes has been transformed through an epistemological shift. Those who comprehend the breadth of epistemic change are most likely to succeed in their strategic endeavors.

The RMA and Synergetic Cocktails

> Chaos has no plural.[4]
>
> *Carlos Fuentes,* The Death of Artemio Cruz

RMAs necessarily entail a rupture, which historically has ranged from a single technological shift, as was the case with the development of the longbow arrow or nuclear weapons, to a sweeping change involving numerous factors. The shift to modernity represented such a change, ushering in the modern nation-state, patriotism, a national army, capitalism, and philosophical notions emphasizing linear progress, balance, objectivity, absolute truth, and so on. Thus, the RMA associated with modernity moved in step with broad changes in numerous realms. We have seen this to be the case with respect to the recent RMA, which has been linked to the shift from modern to postmodern security. This concert of strategic changes has a synergetic effect that is greater than the sum of individual alterations. This synergetic cocktail can assist in explaining strategic change in both Colombia and Mexico.

Asymmetry

In the Colombian case, newfangled asymmetric warfare has been related to numerous features of the RMA. Shifts in political economy are perhaps chief among these. The advent of potent paramilitary forces in the country is a reflection of a rupture in the political economy involving the burgeoning industry of narcotrafficking in the context of economic globalization. The formation of paramilitary forces, in some ways, has provided a huge strategic benefit for both the U.S. and Colombian governments in the war against the FARC and the ELN. The use of paramilitary forces represents a development meant to shift asymmetric war to a more symmetric field by fighting like with like—FARC guerrillas versus AUC paramilitaries. This has eased the state's direct burden in attempts to contain leftist insurgents. But the benefits of the paramilitaries to the U.S. and Colombian governments have been balanced by some formidable costs. Most notable in this regard has been the implosion by early 2007 of the state's attempts to legitimize the paramilitaries through the "demobilization" process. It has entailed mounting and irrefutable evidence of the paramilitary's substantial influence over political, military, economic, and other aspects of Colombian society, and may wind up smearing and seriously debilitating the Uribe government.[5] It is not an easy task to legitimize a criminal group that has engaged in systematic terror against progressive social forces, and it is even harder to do so when evidence points to the collusion between these criminal-terrorists and the state. Overall, in terms of asymmetric warfare and its synergetic relation to other elements of the RMA such as shifts in political economy, we have observed that illicit transnational capital and private armed forces have worked in collusion with the interests of the Colombian and U.S. governments in an attempt to rout the Left.

Let us further pursue the synergetic relationship between asymmetry and other elements of the RMA. Beyond the utility of paramilitary forces for the state in the context of asymmetric warfare, strategic restructuring plans such as Plan Patriota have attempted to renovate the army from its dinosaur status through alterations including smaller and cellular units, and a capacity for rapid action twinned with high mobility in multiterrain environments. This restructuring of the Colombian military has been accomplished in part through a new wave of warfare privatization—the creation of private military and security corporations together with the use of neomercenaries. Part of asymmetric warfare in Colombia has been manifested in the classic cat-and-mouse game of guerrilla warfare in the countryside. A slew of surveillance devices—ranging from high-tech radar systems, to highway panic buttons, to toll-free report-a-guerrilla telephone lines—have been put in place by the state to better cope with asymmetry. The notion of complexity also is involved, to the extent that the result of state actions has not always been positive or even anticipated—for example, pressure in the countryside temporarily pushed the FARC toward urban warfare. Further, the FARC has engaged in nonlinear warfare since at least 2003. The point is that asymmetric warfare cannot really be discussed without reference to the privatization of warfare, newfangled organization, ultra-surveillance, complexity, and the synergetic effect of this mixture.

Mexico has been host to asymmetric warfare in both its traditional and newfangled manifestations. The plight of the EPR's struggle reveals how vulnerable traditional guerrilla groups are to the forces of the Mexican government. Recast as terrorists by a state that itself has admitted unleashing terror against geographically remote social forces, this discourse has framed the EPR rebels as "bad" guerrillas whose actions justify whatever repressive measures the state adopts to rout them out. By contrast, the Mexican government has proven to be considerably weaker vis-à-vis postmodern resistance groups such as the EZLN in Chiapas and the PCs in Guerrero. In the case of the Zapatistas, asymmetric warfare has been employed alongside netwar, identity politics, a rebuke of neoliberal political economy, the reclamation of social space through Caracoles, and a creative use of surveillance from the bottom up through info-war. Guerrero's PCs have quietly reclaimed strategic space through the creation of a largely indigenous police force that appears to have succeeded in substantially reducing crime. Though armed they do not represent an armed insurgency. While the government views them as usurping its sovereign authority, PCs have established genuine consensual power through a strong record of providing legitimate security to social forces weary of official police corruption and military repression. Here we have a synergetic combination of elements of the RMA such as asymmetry, surveillance, network organization, identity politics, and redefined space.

Complexity

Besides asymmetry, complexity as a feature of the RMA is highly apparent with regard to the Colombian imbroglio. For example, the fumigation of hundreds of thousands of hectares in Colombia may have far-reaching and as yet unknown ecological consequences. Because it kills all vegetation on the ground, fumigation alters a complex ecosystem. Further, fumigation takes the livelihood away from thousands of peasants, who have grown desperate in the context of a devastating agricultural crisis brought on in part by economic globalization and the subsidization of Northern crops. This dynamic has pushed peasants into the arms of the FARC, thus enhancing the power of the state's enemies.

It is astonishing that despite the $5 billion price tag of PC, much of which has been devoted to the drug war, between 2002 and 2006 the Pentagon has reduced by 62 percent its surveillance over the Caribbean and Pacific Ocean routes that are commonly used to smuggle Colombian cocaine and heroin. Aerial patrol hours of favorite smuggling routes fell from 6,062 hours in fiscal year 2002 to 1,435 in 2005 and 2,296 in 2006. Beyond that, the navy has reduced by about one-third its patrol boats in the region.[6] These sharp reductions are rooted in the overstretch experienced by the U.S. armed forces since the invasions of Afghanistan and Iraq. All this adds yet another wrinkle to the litany of failures associated with the U.S.-sponsored drug war, and is illustrative of the complex effects emanating from Washington's preoccupation with Iraq in a fashion that has worked at cross purposes to America's most expensive military endeavor outside the Middle East.

The use of contractors demonstrates the relation between complexity and privatization. The line of legal authority blurs when human rights abuses are allegedly propagated by contractors, as evidenced by the bombing of innocent civilians in Colombia by teams involving contractors. Further, complexity is linked to the periodic speculative chaos associated with economic globalization, as manifested in Colombia's volatile stock market and in the country's roller-coaster oil sector. With regard to its ties to epistemology, we have noted that Colombia hosts an array of strategic complications that spring from the country's mix of premodern, modern, and postmodern structures. More broadly, there is an exponential strategic effect among complex elements associated with the RMA, such as narcotrafficking, fumigation, ecocide, the strategic alignment of peasants, the privatization of warfare, globalized economic speculation, epistemological fissures, and so on.

We have noted a plethora of manifestations of complexity in the Mexican case. One notable phenomenon in this regard is the contradictory tendency associated with Nafta-related economic integration versus American post-9/11 security concerns that have spelled a militarized border. Caught in the middle are Mexican migrants. Their hard but illicit work has been vital to the survival of U.S. businesses in an ever more competitive environment. Thus, this reflects a blend of features related to the RMA, including the war on terror, economic regionalization or globalization, and shifting strategic space. The political economy of narcotrafficking represents another obvious and multifaceted case of complexity. It has spawned environmental implications through uncontrollable forest fires in southern Mexico, it has promoted sharp reductions in tourism especially in Guerrero, and it has resulted in government corruption and the debilitation of the state, among other deleterious effects. Yet another prominent example of strategic complexity concerns the fact that the government is potentially being watched by networks of global civil society in instances when it launches repression. This has created a perplexing problem for the state, especially in cases where social forces harness strategies involving info-war, netwar, the adoption of identity politics, and so on. Social forces can enjoy strategic success by appreciating the exponential effect of combining various features of the RMA.

Fear and Terror

Fear and terror represent traditional staples of warfare, but they have been amplified considerably during the current RMA. While the Colombian case in particular demonstrates how very traditional these elements are, what is relatively new in the country is the refashioned discourse on terror since 9/11 and its impact of recasting leftist guerrillas into one of the world's leading terrorist organizations. Further, it is no longer the Liberals and Conservatives who use fear and terror against one another, as they did for more than a century. Rather, there has been a new context featuring the employment of fear and terror by an assortment of privately financed insurgents in addition to the state itself. Narco-terrorism has been perpetuated by varied insurgent and paramilitary forces, thereby denigrating claims to morality by either the right or the left. Importantly, the instruments of fear and terror have often been used in profit-making schemes linked to Northern consumption, especially regarding drugs and oil.

Further, with regard to synergetic relation between terror and other elements of the RMA, it has been noted that labor unions have been the targets of terror unleashed by paramilitary forces and perhaps the state due to issues associated with privatization, economic inequity, and weighty charges of exploitation. The legitimacy of the state and insurgents has been debilitated in the process. Political exclusion has been accentuated, as free expression of ideas has been met with terror. This kind of atmosphere has seriously stifled the development of civil society and has inhibited the expression of identity politics as a highly significant strategic factor as it has been, for example, in the Mexican case. We also observed that in Colombia the strategic instrument of terror has resulted in mass displacement.

Here, then, we observe the synergetic relation of fear and terror with elements of crime, privatization, insurgency, displacement, and with the stifling of civil society and identity politics.

We have observed that the War on Terror has played both ways for the Mexican government. On the one hand, Mexico has been humiliated by the intended creation of a new iron curtain on its frontier with the United States, because Mexico had hoped that Nafta would erase and not fortify distinctions between the two countries. On the other hand, the discourse on terror has helped the government strategically vis-à-vis traditional insurgents such as the EPR. While the release of a government report that documented state terror in Guerrero in the 1960s and 1970s was meant to have the effect of reducing future prospects of state terror, it is remarkable that the government has indeed resorted to acts of terror in its attempt to weaken the EPR as recently as the late 1990s and perhaps later. Some Mexican resistance movements have succeeded by resisting the trap of resorting to organized violence, a snare that can cast these groups as terrorists and thus serve to justify highly repressive measures against them by the government. Finally, the government and civil society have found themselves to be increasingly victimized by narco-terror. Thus, the broad strategic theme of terror in Mexico has worked in concert with other elements of the RMA, such as shifting strategic space, the political spin of discourses, the blurring of crime and warfare, and so on.

Surveillance and Intelligence

Ultra-surveillance employed by the state and the creative use of intelligence-connectivity by resistance forces are signature features of the RMA. Surveillance has been used by the state, the AUC, and the FARC in ways that have sometimes balanced the power of one another. From the perspective of the government, its use of surveillance largely has been linked to U.S. technological wizardry contained in PC. The FARC and the AUC have relied on criminal loot to finance elaborate surveillance and infiltration schemes vis-à-vis government offices. Social forces in Colombia have been victimized by the use of surveillance on the part of the state and the AUC, as we have seen with respect to Colombian labor unions and NGOs. Although there have been increasing glimmers of hope, in general an underdeveloped Colombian civil society has not yet achieved the unity or wherewithal necessary to use surveillance in its favor by attracting in a sustained way the global gaze. This trend has been augmented by the conspicuous absence of a critical press in Colombia, where major media have been closely tied to the government. This stands in contrast to Mexico's *La Jornada, El Sur,* and other critical media that have supported and nurtured the development of civil society in that country. Finally, neoliberal surveillance has worked in tandem with the interests of the United States and TNCs, the Colombian government, and the AUC. The use of ultra-surveillance, then, has worked in harmony with a wide assortment of features associated with the RMA.

The Mexican case provides another panorama of the synergetic relation between surveillance and various elements of the RMA. Mexico has exemplified how neoliberal discipline and economic globalization are strongly supported by surveillance. Sometimes under the guise of "transparency," the phenomenon of surveillance enables foreign investors to withdraw billions of dollars in a matter of nanoseconds if economic fundamentals are perceived to be off-kilter or if serious security problems are imagined to be on the horizon. Surveillance also has been a key element at the increasingly militarized frontier, one meant to reinforce the division between the rich North and the poor South. While in the Colombian case we focused on displacement or internal migration due to warfare, in Mexico we witnessed displacement in the form of a massive exodus of Latin American economic migrants to the United States. A new wrinkle in border surveillance concerns the prospect of privatized security camps located just north of the border and run by the private military corporation Blackwater USA, a development that has alarmed Mexico.[7] Here we note the synergy between privatized warfare, surveillance, trade bloc security, and migration-displacement.

Beyond these examples, it was observed that the Mexican government has employed surveillance against assorted insurgent and resistance groups that have protested against neoliberalism, political exclusion, *caciqueism*, repression, and so on. But ultra-surveillance also can be used to further the power and interests of progressive social forces, as the cases of the Zapatistas and assorted NGOs clearly demonstrate. Nafta membership has decreased the privacy that the Mexican state once enjoyed and that facilitated its ability to unleash unspeakable repression without the threat of considerable global rebuke—as exemplified by the student massacre of 1968. Under the current RMA the challenge has been for social forces to shine the global gaze upon their relations with the state to secure them from repression and to advance their political cause. While devices such as the radio can unite local social forces, as demonstrated by Oaxaca's APPO movement, often more sophisticated equipment and knowledge are required to attract global attention. This is a complicated task, one that involves computer savvy, funding for electronic equipment, as well as an ideological and epistemological sensibility for the global audience one is trying to reach. Overall, the utility of surveillance is variegated and is synergetically related to other features of the RMA such as political economy, asymmetry, organization, and complexity.

Political Economy

It has been underscored that significant economic changes have accompanied strategic transformations in Colombia. Globalization and neoliberal politics have framed many of the country's strategic concerns, ones that have been related to both the criminal sector and the legitimate economy. With respect to the illicit economy, porous borders associated with globalization have combined with the effects of debt restructuring and unfettered free trade to fortify narcotrafficking—an industry fueled by consumption on the part of Northerners who are bombarded on a daily basis with advertisements that promote a culture of (legal) drug use. Further, persistent inequity and exploitation in Colombia's legal economy have underpinned asymmetric warfare and terror. Overall, various elements of postmodern security associated with the RMA in Colombia have been closely tied in a synergetic fashion to neoliberalism and economic globalization.

The same is true for Mexico. For example, Mexico's integration with the United States and Canada under Nafta has meant a profound shift from national to regional security, and greater control over Mexican security by transnational and U.S. interests. Issues such as failed neoliberalism, economic inequity, and the drive to illicit migration have underpinned asymmetric warfare. They have been the common demons identified by a wide variety of resistance movements and insurgents. Further, political economy concerns have framed issues associated with complexity, such as those tied to narcotrafficking, the redefinition of political space at the frontier, and so on. Instruments of fear and terror have been deployed in battles involving political economy, and have been used by the government against social movements opposing the economic shifts that have occurred since the debt crisis of 1982. This is not to suggest that elements of political economy play a determining role with regard to the RMA, but that key shifts linked to the postmodern political economy have interacted in a dynamic way with other ingredients of the RMA's synergetic cocktail.

Identity Politics

Identity politics and related organizational change is a fundamental feature of the RMA, one that assists in explaining strategic change in Colombia and Mexico. Shifts in this regard have occurred for both the state and selected social forces. In Colombia, identity politics has been important in some notable cases, though not to the extent of its rather widespread expression in Mexico. The manifestations of identity politics have been clear in Colombia with respect to displacement and its effects on various populations, such as women, children, indigenous, and blacks. Ethnic identity has been significant in armed clashes over corporate projects and related strategic crises in various parts of the country—from the campaign by the black population of Chocó against the encroachment of their land by transnational palm plantations, to indigenous protests against oil extraction by TNCs in northeastern Colombia. Like other components of civil society, however, organized identity politics often has been muted by ferocious and exclusionary policies on the part of insurgents, paramilitary forces, and sometimes the state. Despite the context of general insecurity for civil society, Colombia's identity politics has been strengthening slowly, especially in cases where international networks can assist their endeavors. Identity politics in Colombia, then, is related to asymmetric war, privatized war, and various elements of political economy.

Turning to Mexico, its entrance into Nafta meant the end of a political culture that exalted "Revolutionary Nationalism" as well as Mexico's peasant and indigenous origins. The Mexican state has struggled to achieve an identity linked to first world status. Paradoxically, many social forces, especially in southern Mexico, where hegemony wears thin, have responded with the concoction of an identity politics that combines a "return to origin" with postmodern sensibilities of rainbow politics. Through this prism, a celebration of Mexico's indigenous past and communal land has been combined with a broad-based politics that encompasses the interests of the marginalized—ones involving issues of gender, ethnicity, ecology, and anti-neoliberalism. Strategies of ultra-surveillance used by social forces often convey the message of identity politics, as they have for resistance movements in Guerrero, Oaxaca, and Chiapas. Thus, there is an exponential effect among combined features of the RMA such as identity politics, surveillance, organization, and the forces of economic regionalism and globalization.

Lessons Learned

> Power corresponds to the human ability not just to act but to act in concert.[8]
>
> *Hannah Arendt,* On Violence

Perhaps the most radical question one can ask with regard to the RMA is "A revolution for whom?" The interests of the state and insurgent forces are crucial and have been considered here from a number of angles. Beyond that, a principal purpose of this study has been to include an analysis of the often overlooked elements of civil society and the majority population of the poor in order to explore how they can empower themselves through navigating elements of the RMA, rather than being its victims. Among the key avenues to success are strategies that include adaptability and flexibility, regional and global networking, unity, the instant dissemination of information, epistemological sensibilities, nonbelligerent reclamations of space, and taking it to the streets. Let us proceed to examine the importance of the RMA for both the state and civil society, beginning with the former.

The RMA as manifested in PC has yielded a mixed bag of successes and failures for the governments of both the United States and Colombia. PC has served their interests in the short term, but seems prone to ultimate failure due to its neglect of the material and political interests of the country's majority population. With regard to short-term successes, the plan has helped restore some semblance of order in the country after the mounting chaos that prevailed in the 1990s. For example, between 2002 and 2006, homicides have been reduced by about one-third. Much of this has to do with the attempt by the paramilitaries to exhibit relatively good behavior in order to be politically legitimized through the demobilization process. During that same period, PC has also helped reduce by about 60 percent the number of attacks by leftist insurgents, especially against the country's petroleum installations and other infrastructure.

Elements of the RMA such as ultra-surveillance, privatized war through the use of neomercenaries and assorted military corporations, and a newfound adaptability and flexibility exemplified through organizational change of the armed forces have all served to strengthen the state. So, too, has the government's manipulation of the discourse on terror—a feat that simultaneously demonized the FARC and the ELN while attempting to permit the legitimization of right-wing terrorists under the banner of the AUC. It has been emphasized that the Colombian population has been hungry for a taste of everyday order, and there has been spotty progress in this direction. By 2007 there was a sense that leftist military forces such as the FARC and especially the ELN were on the defensive, at least in comparison to their ascendant position in the late 1990s. There even existed a glimmer of hope that a long-sought-after peace accord between the government and the ELN might be achievable. Finally, the success of the country's Polo Party in 2006, which displaced the Liberals as the official opposition party, represents a highly positive development for Colombia's budding civil society. Thus, the tone of a carnage-ridden power grab that blighted the 1990s has given way to a relatively calmer period during the first decade of the new millennium.

This more orderly environment, together with vague promises of a future that features diminished warfare, has underpinned support for the right-wing Uribe government. Much like the initial popularity of Peru's Alberto Fujimori, which appeared in the wake of Sendero Luminoso's defeat in the early 1990s, the popularity of Uribe demonstrates that society wants order as a first step to political and economic development. This new era of militarized order under the RMA has also resulted in pronounced macroeconomic growth in the country, especially growth linked to foreign investment. Uribe's business-friendly neoliberal policies have attracted transnational capital, and this trend has been augmented by the military protection of foreign investment afforded by PC. Nowhere is this clearer than in the petroleum sector, where key pipelines have at times been protected by U.S. Marines in addition to the regular and fortified patrols by U.S.-trained Colombian forces. Petroleum production rose 9.4 percent from 2005 to 2006, leading to the industry's best output in the country since 1999. Further, some key petroleum corporations have chosen to invest in Colombia rather than in Venezuela, where the "government take" can be 60 percent or more under the policies of Hugo Chávez, especially in the highly promising Orinoco fields.[9] Thus, the modest trend toward a more orderly environment and the concoction of TNC-friendly policies have combined to fortify neoliberalism in Colombia, in contrast to the burnout of this model almost everywhere else in Latin America. What remains to be seen is the extent to which foreign-led macroeconomic growth trickles down to the majority, a phenomenon for which there is little evidence at the present juncture.

Fujimori's case showed that the provision of order is just the first step to achieving political popularity. Once order is established, or at least accentuated as it has been in the case of Colombia, there is likely to be a sharp rise in public demands for government legitimacy, truer democratic structures, reduced inequity, greater political plurality, and so on.[10] It is worth underscoring that these are precisely the issues that underpinned the formation of guerrilla movements in the first place—although these crucial themes have been obscured due to the FARC's and the ELN's downward spiral into a military machine fueled by criminal largesse. Overall, as the crash and burn of Peru's Fujimori demonstrated, the provision of initial order must be followed by sweeping political and economic reforms. It is not yet clear whether Uribe has learned from Peru's experience. The Colombian government's penchant for the political exclusion of civil society, severe repression aimed at those who rightfully complain of excruciating inequity and exploitation, and alarming criminal links between the Colombian state, paramilitary forces, and TNCs, are all worrisome signs with regard to positive political development in Colombia. To be successful an RMA must serve the interests of the majority population rather than just the narrow objectives of the state and TNCs.

Although there exist some important and increasingly frequent exceptions, nonarmed social forces in Colombia generally have not yet fared well under the RMA. Leftist belligerents have continued to foment a climate of fear and even terror that is not propitious for free speech and political organization. Insurgents on both the left and the right have perpetuated massive displacement. Right-wing paramilitaries have been particularly threatening to the development of peaceful social forces. They have been the leading purveyor of assassinations of and threats to members of labor unions and other progressive elements of civil society. For example, the leader of Colombia's USO labor union in Cartagena, which represents petroleum workers, was forced to flee the country along with his family in the summer of 2006 due to death threats by paramilitaries.[11] Beyond their effect on labor, which arguably has remained the most organized component of the country's fledgling civil society, the paramilitary influence also has stifled the development of critical NGOs. The situation has been exacerbated by the exclusionary policies of the government, which, for example, has not included civil society in any peace negotiations with insurgents. Overall, the key to development of social forces in Colombia is to organize en masse, but the climate of fear and exclusion has not subsided sufficiently for that to occur in a sustained fashion.

Despite this, there has been some progress among social forces in Colombia. Importantly, the significance of using ultra-surveillance tools is appreciated by key players. A leader of the CUT, a major labor union in the country, acknowledged that attracting "international attention is key to the security of labor unions."[12] It was observed that labor organizations have managed to convince the International Labor Organization to open an office in Bogotá. This example of global networking will help shine the global gaze on the appalling repression union members have faced in Colombia. Turning to the other major social force considered in this study, the displaced have had a strong advocate in NGOs such as Codhes. That group also has used surveillance and global networking to promote the interests of those it serves. But the fact that the overwhelming majority of displaced do not register with NGOs or government offices, together with their failure to organize en masse, demonstrates that considerable work needs to be done.

* * *

Turning to Mexico, let us begin by tallying the strategic wins and losses for the state with regard to the RMA. On the positive side of the ledger, the government has done relatively well with regard to the use of surveillance and intelligence vis-à-vis insurgents, as the case of the EPR clearly demonstrates. In a similar vein, the government has aptly managed asymmetric warfare against traditional groups such as the EPR and its offshoots largely because these organizations failed to keep pace with strategic change. Related to these issues, the state shrewdly manipulated the issue of terror with the construction of a discourse that pitted "good" but militarily powerless guerrillas against "bad" insurgents who rely on armed force.

The most profound weakness of the Mexican government has been its failure to address the root causes that generate resistance movements in the first place. That is, the EZLN, the EPR and its offshoots, and Guerrero's PCs, in addition to a host of other resistance movements, have all spawned in reaction to the failure of the neoliberal economic model. Anemic performances in the manufacturing, agricultural, and tourist sectors, combined with vulnerabilities in the speculative extractive sector, have set the stage for mass migration and social discontent. In Chapter 6, where we observed the syndrome of "fight or flight," figures from ECLAC were highlighted that suggest that Mexico's poverty rate would increase by almost 50 percent if it were not for remittances. In the context of tightening border security, the safety valve of remittances may face significant decline. Against that backdrop, it is imperative for the conservative regime of Calderón to appreciate fundamental social and economic weaknesses in the neoliberal approach, as have most other major Latin American countries, or else face mounting strategic crises. With more than a quarter of a century down the garden path of neoliberalism, the Mexican state must choose between staying the economic course and a strategic crisis waiting to happen.

Given its high level of integration with the United States, the Mexican government has found itself perhaps more vulnerable than any other Latin American country to the shifting winds of American politics. For example, the ambitious and progressive migration policies to which former president Fox and his foreign minister Jorge Castañeda aspired never reached fruition due to U.S. intransigence post 9/11. Besides having to cope with the United States calling the shots on the migration issue, Mexico has also faced a mounting narcotrafficking crisis due largely to American consumption and to a refusal by Washington to consider the legalization or regulation of now-illicit drugs—a move that would completely disempower the terror-prone syndicates that feed off the trade. While the government of Calderón has attempted to curtail the military power of narcotraffickers by sending nearly 7,000 Mexican military troops to his home state of Michoacan in December 2006, as well as 3,000 troops to Tijuana and 7,600 troops to Guerrero in early 2007, this approach is not likely to yield a significant impact because the root causes for the existence of this multibillion dollar industry remain unaddressed. For Mexico, then, a big part of the notion of complexity concerns grappling with predicaments generated largely by the United States.

Elements of civil society have experienced varying degrees of success working with components of the RMA. Perhaps the simplest lesson here is for social forces to appreciate that an RMA has indeed occurred, and to explore creatively its manifestations. This has been accomplished on the part of Chiapas' EZLN and its wide band of supports, who have relied on information warfare, netwar, identity politics, and so forth. While NGOs and resistance movements have thrived in Chiapas against the backdrop of the Zapatista uprising, such groups are beginning to come of age in other parts of southern Mexico, such as Guerrero and Oaxaca. An unsung hero in this regard is the Policías Comunitarias in Guerrero, which represents a highly successful experiment in biopolitics that has provided a genuine regime of human security to a population previously victimized by exclusion and repression. But deep challenges remain for social forces in Guerrero and Oaxaca, including threats against unity, a weaker history of community organization than has been the case in Chiapas, the geographic dispersion of social forces, and the expense and expertise required to use ultra-surveillance as a means of seeking national and global support for localized crises. Groups such as Tlachinollan in Guerrero and the APPO in Oaxaca have hinted that these challenges can indeed be overcome.

Strategic losers in Mexico include groups that have relied on traditional approaches without appreciating transformations linked to the RMA. Exhibit A in this regard is the EPR and its progeny. While attempting to represent the worthy cause of a population in southern Mexico victimized by repression, political exclusion, and economic marginalization, these insurgents failed those that they attempted to assist by adopting a strategy out of synch with developments associated with the RMA. They relied on the ideology, rhetoric, and strategy of 1960s guerrillas in Guerrero, who themselves were ultimately unsuccessful and were victimized by a government Dirty War that unleashed terror and genocide. The EPR and its splinter groups ineptly struggled to achieve the status of a "regular" guerrilla army rather than pursue clever manifestations of irregular and asymmetric warfare. They were not adept at using the electronic medium linked to information warfare and ultra-surveillance. They could not concoct a message that could be seized upon by global civil society, because they were not attuned to ideological and epistemological changes that had occurred outside their realm. Finally, they relied on force rather than on the power of ideas, and fell victim to the government's attempt to portray them as terrorists. The EPR in the summer of 2007 detonated a series of ten bombs on natural gas pipelines operated by PEMEX in Veracruz and elsewhere. In a desperate attempt to push Mexican authorities to release two of its leaders it claimed the government held in captivity, the rebels demonstrated their renewed willingness to use considerable force. The surprise bombings, which will cost millions of dollars in pipeline repairs and which temporarily shut major businesses relying on the gas, indicate a certain sophistication since some of them were detonated simultaneously. But the rebels' strategy was naive to the extreme, since Mexican authorities almost certainly will not give in to any rebel demands. Instead, such actions are likely to increase government repression in EPR hotspots such as Guerrero and Oaxaca, thereby prompting a more repressive environment for the social forces the rebels claim to represent. This represents another example of the "Colombianization" of Mexico, where insurgents have become experts at destruction rather than participants in the creation of a better society for the majority population.

Final Thoughts

> Discontinuity—the fact that within the space of a few years a culture sometimes ceases to think as it had been thinking up till then and begins to think other things in a new way.[13]
>
> *Michel Foucault,* The Order of Things

A truly revolutionary development in the Latin American context is that since the late 1990s social revolution has transpired through democratic means rather than through warfare. Democracy has provided revolutionary results in countries such as Venezuela, Bolivia, Ecuador, Argentina, Nicaragua, and others. The success or failure of these revolutions remains to be determined, though it seems clear that violent and criminally oriented guerrilla groups have grown increasingly isolated and outmoded. It is crucial to emphasize that democratic trends in Latin America owe much to the United States. Neoliberal restructuring since the 1980s has witnessed the parallel development of steadily more elections that seem to reflect real democracy. That is, while the neoliberal model arguably has harmed the material interests of the majority population in the region, in many cases democratic structures have allowed them a peaceful means by which to opt out of that failed model of development. While some in the United States may shudder at such thoughts, and dismiss the move to the left in Latin America as another example of outdated Marxism, there are other relevant historical and philosophical references from which Americans may take heed. Thucydides, for example, attributed Athens' loss in the Peloponnesian War partly to Athens' exploitative relation with its colonies, which sided with Sparta when the first opportunity presented itself.[14] Great and enduring leadership relies on consent and fulfilling the interests of those who are led. That is the stuff of which the heroes of the current RMA are made.

Notes

Chapter 1

1. Miguel Cervantes, *Don Quixote* (New York: Oxford University Press, 1999), p. 71.
2. For an excellent historical discussion of the RMA, see Andrew Latham, *Understanding the RMA: Braudelian Insights into the Transformation of Warfare* (Geneva: Programme for Strategic Studies and International Security Studies, 1999); and Azar Gat, *The Origins of Military Thought from Enlightenment to Clausewitz* (Oxford: Oxford University Press, 1989).
3. See Andrew F. Krepinevich, "Calvary to Computer: The Patterns of Military Revolutions," *The National Interest,* #37, Fall 1994, p. 33.
4. See Michael Dartness, "Insurgency Online," *Small Wars and Insurgencies,* vol. 10, #3, Winter 1999, p. 129.
5. Thucydides, *History of the Peloponnesian War*, ed. Robert Strassler (New York: Free Press, 1996), p. 43.
6. Michel Foucault, *The Order of Things: An Archaeology of the Human Sciences* (New York: Vintage, 1970), p. 387.
7. Michel Foucault, *Remarks on Marx* (New York: Semiotext(e), 1991).
8. Homer, *Iliad* (New York: Barnes and Noble, 1995), pp. 31–33 and 43.
9. Ibid., p. 8.
10. Ibid., p. 9.
11. Ibid., p. 17.
12. Thucydides, *History of the Peloponnesian War,* p. 3.
13. See, for example, Lothar Ibugger, "The Revolution in Military Affairs," NATO Parliamentary Assembly Paper, November 1998, pp. 5–10; and Latham, *Understanding the RMA*.
14. See Elizabeth Brook, "Information Warfare Pioneers Take Top Pentagon Positions," *National Defense Magazine,* January 2002, www.nationaldefensemagazine.org, viewed 7 January 2002.
15. See the excellent discussion by Michael Dillon and Julian Reid, "Global Liberal Governance: Biopolitics, Security and War," *Millennium,* vol. 30, #1, 2004, p. 44.
16. For a good discussion of asymmetric warfare and the RMA, see Jeremy Black, "War and Strategy in the 21st Century," *Orbis,* vol. 46, #1, Winter 2002, pp. 137–144.
17. See Dirk Kruijt and Kees Koonings, "Introduction: Violence and Fear in Latin America," in *Societies of Fear: The Legacy of Civil War, Violence and Terror in Latin America,* ed. Kruijt and Koonings (London: Zed, 1999), p. 11.
18. See, for example, Thomas Adams, "The Real Military Revolution," *Parameters,* vol. 30, #3, Autumn 2000, pp 54–65, especially p. 54.
19. Michel Foucault, *History of Sexuality* (New York: Vintage, 1978), p. 86.
20. See Alejandro Portes and Kelly Hoffman, "Latin American Class Structures: Their

Composition and Change during the Neoliberal Era," *Latin American Research Review,* vol. 38, #1, 2003, pp. 1–35.
21. See R.B.J. Walker, *Inside/Outside: International Relations as Political Theory* (Cambridge: Cambridge University Press, 1993).
22. Thucydides, *History of the Peloponnesian War,* p. 112.
23. Ibid., p. 7.
24. For example, in relation to the attempted Athenian conquest of Syracuse, Thucydides observes the irony that "[t]hey had come to enslave others, and were departing in fear of being enslaved themselves." In the course of the Peloponnesian War, Athens ultimately lost partly due to the fact that some of its enslaved colonies took any opportunity that presented itself to join ranks with the rival Spartans. Thucydides, *History of the Peloponnesian War,* p. 472.
25. See, for example, Sun Tzu II, *The Lost Art of War* (San Francisco: HarperCollins, 1996), p. 67.
26. See Niccolo Machiavelli, *The Prince* (New York: Mentor, 1980), p. 86. For a discussion of class conflict, see also Machiavelli, *The Discourses* (New York: Penguin, 1970), especially pp. 111–118.
27. Thomas Hobbes, *Leviathan* (New York: Penguin, 1982), p. 192.
28. Ibid., p. 193.
29. Charles Dokubo, "An Army for Rent, Private Military Corporations," *Civil Wars,* vol. 3, #2, Summer 2000, p. 53.
30. Thomas Adams, "The New Mercenaries and the Privatization of Conflict," *Parameters,* vol. 29, #2, p. 104.
31. See MRPI's Web page, www.mpri.com, under "capabilities."
32. An example is AirScan.
33. United Nations Office of the High Commissioner of Human Rights, "Protocol Additional to the Geneva Convention of 12 August 1949," www.unhchr.ch/html/menu3/b/93.htm, viewed 30 April 2004.
34. Tony Lynch, "The Good Mercenary?" *Journal of Political Philosophy,* vol. 8, #2, 2000, p. 133 (see broader article for good overall discussion, pp. 133–153).
35. See Eugene Smith, "The New Condotieri and US Policy: The Privatization of Conflict and Its Implications," *Parameters,* vol. 32, #4, Winter 2002–2003, p. 106.
36. See William Caferro, *Mercenary Companies and the Decline of Siena* (Baltimore: Johns Hopkins University Press, 1998), p. xiii.
37. See Machiavelli, *The Prince,* p. 77.
38. See Guy Arnold, *Mercenaries: The Scourge of the Third World* (New York: St. Martins, 1999), pp. x–xi.
39. See Stephen Ferris and David Keithly, "Outsourcing the Sinews of War: Contractor Logistics," *Military Review,* vol. 81, #5, September-October 2001, pp. 72–83; www.uscac.cac.army.mil/CAC/milreview/English/sepoct01/indexso01.htm, viewed 3 October 2003.
40. Smith, "New Condotieri and US Policy," p. 106.
41. Ferris and Keithly, "Outsourcing the Sinews of War."
42. See Abdel-Fatua and J. Kayode Fayemi, eds., *Mercenaries: An African Security Dilemma* (London: Pluto, 2000), p. 77.
43. See, for example, David Francis, "Mercenary Intervention in Sierra Leone: Providing National Security or International Exploitation?" *Third World Quarterly,* vol. 20, #2, 1999, pp. 319–339; Pratap Chatterjee, "Guarding the Multinationals," *Multinational Monitor,* vol. 19, #3, March 1998, pp. 1–5; and Doug Brooks, "Messiahs or Mercenaries?" *International Peacekeeping,* vol. 7, #4, Winter 2000, pp. 129–144.

44. First quote from Anthony Mockler, *The New Mercenaries* (London: Sidgwick and Jackson, 1985), p. 5; second quote from Carlos Escude, *Mercenarios del fin del Milenio* (Buenos Aires: Editorial Belgrano, 1999), p. 15.
45. See Linda Robinson, "America's Secret Armies," *US News and World Report,* vol. 133, #17, November 2002, pp. 5–9, www.usnews.com.
46. See, for example, William Avilés, *Global Capitalism, Democracy, and Civil-Military Relations in Colombia* (Albany: SUNY, 2006), pp. 106–107.
47. A UN resolution was passed when the United States showed no inclination to heed the World Court's ruling of 27 June 1986 that U.S. support for the Contras was illegal. For a discussion of this, see James Rochlin, *Discovering the Americas: The Evolution of Canadian Foreign Policy towards Latin America* (Vancouver, Canada: UBC Press, 1994), pp. 157–158.
48. See, for example, Roberto Bardini, *Monjes, mercenarias y mercaderes: La red secreta de apoyo a los contras* (Mexico City: Mex-Sur Editorial, 1988), pp. 7 and 26–27.
49. As reported in *El Sur,* 23 May 2006.
50. For an excellent feminist perspective to the notion of consumption, see V. Spike Peterson, *A Critical Rewriting of Global Political Economy* (London: Routledge, 2003).
51. See Foucault, *History of Sexuality,* vol. 1, *An Introduction,* p. 81.
52. For a good overview of information warfare, see Timothy Shimeall et al., "Countering Cyber War," *NATO Review,* vol. 49, #4, Winter 2001, pp. 16–18.
53. For a broad discussion of the RMA and its relation to market forces, see Steven Metz, "The Next Twist of the RMA," *Parameters,* vol. 30, #3, Autumn 2000, pp. 40–53.
54. See, for example, Terry Ryan, "Committing to a Future of ISR Supremacy," *American Intelligence Journal,* vol. 21, #1, Spring 2002, pp. 7–16. With regard to the distinction between reconnaissance and surveillance, see pp. 8–9.
55. See Jeremy Bentham, *The Panopticon Writings,* ed. M. Bozovic (London: Verso, 1995), quotes from pp. 34 and 43, respectively.
56. See Michel Foucault, *Discipline and Punish: The Birth of the Prison* (New York: Vintage, 1977), p. 201.
57. See Ryan, "Committing to a Future of ISR Supremacy," pp. 7–16.
58. For a good discussion of the importance of human intelligence, see Richard Betts, "Fixing Intelligence," *Foreign Affairs,* vol. 18, #1, January 2002, pp. 43–60.
59. Homer, *Iliad,* see, for example, pp. 144 and 149.
60. See chapter 13 of Sun Tzu, *Art of War,* pp. 168–172.
61. Ibid., p. 168.
62. Miyamoto Musashi, *The Book of Five Rings* (New York: Kodansha, 2001), p. 79.
63. See chapter 6, book 1 of Carl von Clausewitz, *On War,* ed. Peter Paret (Princeton, NJ: Princeton University Press, 1976), pp. 117–118.
64. See Napoleon, *How to Make War* (New York: Ediciones La Calavera, 1998), p. 15.
65. See Chris Demchak, "Technology's Knowledge Burden, the RMA and the IDF: Organizing the Hypertext Organization for Future 'Wars of Disruption'?" *Journal of Strategic Studies,* vol. 24, #2, 2001, pp. 77–147; Patrick Morgan, "The Impact of the Revolution in Military Affairs," *Journal of Strategic Studies,* vol. 23, #1, 2000, pp. 132–163; and Robert Tomes, "Relearning Counterinsurgency Warfare," *Parameters,* vol. 19, #1, 2004, pp. 16–28.
66. See Bruce Berkowitz, "Better Ways to Fix US Intelligence," *Orbis,* vol. 45, #4, Fall 2001, pp. 609–620.
67. See Thomas Quedensley, "The Commercial Satellite Multispectral Imagery Threat," *American Intelligence Journal,* vol. 21, #2, Spring 2002, pp. 33–36.
68. Ibid.

69. For further discussion of surveillance and asymmetry, see David Lyon, "An Electronic Panopticon? A Sociological Critique of Surveillance Theory," *Sociological Review,* vol. 41, #1, 1993, pp. 653–678.
70. See Robert Gibson, "Space Power, the Revolution in Military Affairs" (Strategic Research Project Paper, U.S. Army War College, Carlisle, PA, 10 April 2001), p. 2.
71. Napoleon, *How to Make War,* p. 11.
72. Antoine Henri de Jomini, *The Art of War* (London: Greenhill, 1996), p. 328.
73. Carlos Fuentes, *The Burning Mirror* (New York: Mariner, 1999), pp. 35–36.
74. See, for example, Dillon and Reid, "Global Liberal Governance"; Richard Ek, "A Revolution in Military Geopolitics?" *Political Geography,* vol. 19, #7, 2000, pp. 841–874; Dartness, "Insurgency Online"; Theodor Galdi, "Revolution in Military Affairs? Competing Concepts, Organizational Responses, Outstanding Issues" (paper presented to the Federation of American Scientists, Washington, DC, 11 December 1995), pp. 11–25.
75. For a general discussion of this, see, for example, Keith Hayward, "The Globalization of Defense Industries," *Survival,* vol. 43, #2, 2002, pp. 115–132; Michael Chinworth, "The RMA: A US Business Perspective," in *Managing the Revolution in Military Affairs,* ed. Ron Matthews and John Treddenick (New York: Palgrave Macmillan, 2001), pp. 133–155.
76. B. H. Liddell Hart, *Strategy* (New York: Meridian, 1991), p. 220.
77. Bert Chapman, "Revolution in Military Affairs," *Bulletin of Bibliography,* vol. 59, #1, 2002, p. 352.
78. Franz Fanon, *The Wretched of the Earth* (New York: Grove Press, 1963), p. 36.
79. Ibid., p. 73.
80. Ibid., p. 94.
81. Hobbes, Leviathan, p. 168.
82. Giambattista Vico, *The New Science of Giambattista Vico* (Ithaca, NY: Cornell University Press, 1976), p. 115.
83. Ibid., p. 336.
84. Ibid.
85. Napoleon, *How to Make War,* p. 1.
86. Foucault, *Discipline and Punish,* pp. 8 and 48.
87. Government terror in Latin America has been defined as "beating, killing, robbing, bombing or other assaults on a civilian population, including relatively unusual items such as forced relocation; (2) beating, torturing, or killing of combatants who have indicated a willingness to surrender; (3) the use of weapons which do not sufficiently discriminate among combatants and others." See Timothy Wickham-Crowley, "Terror and Guerrilla Warfare in Latin America, 1956–1970," *Comparative Studies in Society and History,* vol. 32, #2, April 1990, p. 203.

Chapter 2

1. Ahead of Colombia in 2005 were Iraq, Afghanistan, Israel, Egypt, Pakistan, and Jordan. For additional related information, see Curt Tarnoff and Larry Nowels, Congressional Research Service, *CRS Report for Congress: Foreign Aid; An Introductory Overview of US Programs and Policy* (Washington, DC: Library of Congress, 2004), p. 13.
2. An excellent discussion of Colombia's violence is Malcolm Deas and Fernando Gaitán Daza, *Dos ensayos especulativos sobre la violencia en Colombia* (Bogotá: Tercer Mundo, 1985).

3. Even cannibalism was apparent in the region of present-day Cali and northward. See Frank Safford and Marco Palacios, *Colombia: Fragmented Land, Divided Society* (New York: Oxford University Press, 2002), p. 21.
4. See Paul Oquist, *Violence, Conflict and Politics in Colombia* (New York: Academic Press, 1980), p. 26.
5. See John Coatsworth, "Colombia: Roots of Violence in Colombia," *Revista: Harvard Review of the Americas,* vol. 2, #3, Spring 2003, p. 8.
6. For a broader discussion of this, see Charles Berquist, *Labor in Latin America: Comparative Essays on Chile, Argentina, Venezuela and Colombia* (Palo Alto, CA: Stanford University Press, 1986), p. 293.
7. For a broader discussion of this, see James Rochlin, *Vanguard Revolutionaries in Latin America: Peru, Colombia and Mexico* (Boulder, CO: Lynne Rienner, 2003), chapter 2.
8. See Oquist, *Violence, Conflict and Politics in Colombia,* p. 10.
9. See Eric Hobsbawn, *Rebeldes Primitivos* (Barcelona: Ediciones Ariel, 1980), p. 229.
10. See Consultoría para los Derechos Humanos y el Desplazamiento Forzado (Codhes), *Un Pais Que Huye: Desplazamiento y violencia* (Bogotá: Codhes, 2003), p. 146.
11. This point is developed fully in Nazih Richani's superb work *The Political Economy of War and Peace in Colombia* (Albany: SUNY, 2002). Richani argues convincingly that, especially since the 1980s, subversive groups, criminal syndicates, and the military establishment have reaped high profits from Colombia's special brand of warfare and that this factor perpetuates the war. A similar point is made by Eduardo Pizarro León-gomez in *Una democracia asediada—balance y perspectivas del conflicto armado en Colombia* (Bogotá: Norma, 2004), p. 91.
12. Regarding the use of the word "terror" to describe the Colombian case prior to post-9/11 usage, see Daniel Pecaut, "The Loss of Rights, the Meaning of Experience, and Social Connection: A Consideration of the Internally Displaced in Colombia," *International Journal of Politics, Culture and Society,* vol. 14, #1, 2000, p. 90.
13. Álvaro Mutis, *The Adventures and Misadventures of Maqroll* (New York: New York Review Books, 2002), p. 270.
14. Thomas Hobbes, *Leviathan,* ed. A. P. Martinich (Peterborough, ON: Broadview Press, 2002), p. 95.
15. See Rochlin, *Vanguard Revolutionaries in Latin America,* chapter 4.
16. Present-day Colombia was part of "Gran Colombia" during the Bolivarian period from 1819 to 1830, and was called Nueva Granada from 1830 to 1858.
17. For a good discussion of this, see Carlos Fuentes, *The Burning Mirror: Reflections on Spain and the New World* (New York: Mariner Books, 1999), pp. 255–258.
18. See, for example, David Sowell, *The Early Colombian Labor Movement: Artisans and Politics in Bogotá, 1882–1919* (Philadelphia: Temple University Press, 1992), p. 26.
19. For a broader discussion of this, see Safford and Palacios, *Colombia,* pp. 10, 161, and 259; and Sowell, *Early Colombian Labor Movement,* pp. 2 and 17.
20. See Berquist, *Labor in Latin America,* pp. 275–368.
21. See Sowell, *Early Colombian Labor Movement,* p. 39.
22. Gabriel García Márquez, *Living to Tell the Tale* (New York: Alfred A. Knopf, 2003), p. 303.
23. See Alberto Mayor Mora, *Etica, Trabajo y Productividad en Antioquia* (Bogotá: Tercer Mundo, 1984), p. 447.
24. See, for example, Cecilia Herrera, "City as a Modernizing Paradigm," *Pedagogica Historica,* vol. 49, #1, 2003, pp. 65–69.
25. See Nazih Richani, *Systems of Violence: The Political Economy of War and Peace in Colombia* (Albany: SUNY, 2002), p. 16.

26. For an excellent discussion of this, see Francisco Leal Buitrago, *La Seguridad Nacional a la Deriva: Del Frente Nacional a la Posguerra Fria* (Bogotá: Alphaomega, 2002), especially pp. 2–18.
27. For a broader discussion of this, see Doug Stokes, "Why the End of the Cold War Doesn't Matter: The US War of Terror in Colombia," *Review of International Studies,* vol. 29, #4, October 2002, p. 577.
28. See James Rochlin, *Discovering the Americas: The Evolution of Canadian Foreign Policy towards Latin America* (Vancouver, Canada: UBC Press, 1994), pp. 211–212.
29. See *El Tiempo,* 2 May 2006.
30. Fuerzas Armadas Revolucionarias de Colombia (FARC), "Nuestra Historia," 2004, www.farc-ep.org/aniversario/especial40aniv/textcrono.html, viewed 2 May 2005.
31. FARC, "Las FARC-EP: 30 años de lucha por la paz, democracia y soberanía," May 1994, http://www.analitica.com/biblioteca/farc/30.asp, viewed 2 May 2005.
32. FARC, "Historia de las Conferencias," May 2004, http://farc-ep/org/aniversario/especial40/text04.html, viewed 5 January 2005.
33. FARC, "40 años de lucha por la paz," May 2004, http://www.farc-ep.org/aniversario/especial40aniv/text10.html, viewed 10 January 2005.
34. Ibid.
35. FARC, "Nuestra Historia."
36. Ibid.
37. FARC, "Historia de las Conferencias."
38. FARC, "30 años de las FARC-EP, Texto Completo del Discurso Pronunciado por el Comandante en Jefe de las FARC-EP Manuel Marulanda Vélez," 27 May 1994, http://six.swix.ch/farcep/Nuestra_historia/30_anos_manuel.htm, viewed 3 December 2004.
39. FARC, "Nuestra Historia."
40. Alfredo Rangel, "El Repliegue de las FARC: Derrota o Estratégia?" Fundación Seguridad y Democracia, 2004, p. 13, www.seguridadydemocracia.org, viewed 19 February 2005.
41. Ibid., p. 11.
42. The estimate of 17,000 troops should not be viewed as precise, as discussed earlier in the chapter. This estimate represents a general consensus among analysts.
43. See Ejército de Liberación Nacional (ELN), "Quienes Somos?" www.eln-voces.com, viewed 10 January 2005.
44. Leóngomez, *Una democracia asediada,* p. 101.
45. See Gustavo Almario Salazar, *Historia de los Trabajadores Petroleros* (Bogotá: Ediciones Cedetrabajo, 1984), p. 169.
46. Nicholás Rodríguez Bautista, "Ejército de Liberación Nacional: Una historia de vida," in *ELN: Una historia contada a dos voces,* ed. Carlos Medina Gallego (Bogotá: Rodríguez Quito Editores, 1996), p. 159.
47. In terms of art, Álvaro Mutis discusses the absence of a state presence in much of Colombia and its effects vis-à-vis business and other social forces. See, for example, Mutis, *Adventures and Misadventures of Maqroll,* pp. 72–73.
48. For a broader discussion of this, see Rochlin, *Vanguard Revolutionaries in Latin America,* pp. 104–106.
49. For example, it appears on the cover of the ELN's "Posición del ELN sobre la privatización de ISA-ISAGEN," pamphlet distributed by the group in Colombia, March 2000.
50. ELN, "Quienes Somos?" "Historia," and "Por que luchamos," www.eln-voces.com, viewed 10 February 2004.
51. ELN, "Quienes Somos?" viewed March 2005.
52. ELN, "Correo del Magdalena," no. 115, 18 October 1999, www.eln-voces.com, viewed 7 January 2002.

53. Gabriel García Márquez, *One Hundred Years of Solitude* (New York: Avon, 1970), p. 163.
54. Jeremy McDermott, "FARC and the Paramilitaries Take Over Colombia's Drug Trade," *Janes Intelligence Review,* 1 July 2004, www.janes.com.
55. See Mauricio Romero, *Paramilitares y autodefensas, 1982–2003* (Bogotá: IEPRI, 2003), p. 133.
56. Autodefensas Unidas de Colombia (AUC), "27 años de lucha por la paz," February 2005, www.colombialibre.org/detalle_col.php?banner=editorial&id=10507, viewed 20 March 2005.
57. Defensoria del Pueblo—Colombia, Secretariado Nacional de Pastoral Social Caritas Colombiana, *Antioquia: Nordeste* (Bogotá: Defensoria del Pueblo, 2000), p. 26.
58. Romero, *Paramilitares y autodefensas,* p. 120.
59. See Alejandro Reyes, "Drug Trafficking and the Guerrilla Movement in Colombia," in *Drug Trafficking in the Americas,* ed Bruce Bagley and William Walkers, III (Coral Gables: University of Miami, 1994), p. 125.
60. AUC, "Discurso del Jefe del Estado Mayor de las AUC, Comandante Salvatore Mancuso, en el acto de Instalacion Oficial del Proceso de Negociación entre el Gobierno Nacional y las Autodefensas Unidas de Colombia," July 2004, www.colombialibre.org/ver_imp.php?Varid=6425, viewed 18 October 2005. "Reinstitutionalization" is a term frequently used by the AUC to denote a reconstruction of the state with inclusion of paramilitary groups. See Estado Mayor Negociador AUC, "Comunicado a la Opinión Pública," 12 August 2004, www.colombialibre.org/ver_imp.php?Varid=7118, viewed 18 October 2005.
61. See Rochlin, *Vanguard Revolutionaries in Latin America,* p. 147.
62. See, for example, Romero, *Paramilitares y autodefensas,* pp. 30 and 40.
63. Gobierno de Colombia, Ministerio de Defensa, "Los Grupos ilegales de Autodefensa en Colombia," December 2000, www.derechoshumanos.gov.co/observatorio/04_publicaciones/, viewed 28 April 2004.
64. Márquez, *Living to Tell the Tale,* p. 84.
65. For a broader discussion of this, see Alexander Robin, "Colombian Trade Union Leader Patricia Buritica Rallies Opposition to Plan Colombia," *Guild Notes,* vol. 24, #2, Summer 2000, p. 2.
66. See Fernán González, *Pasado y Presente del Sindicalismo Colombiano* (Bogotá: Centro de Investigación y Accion Social, 1975) p. 5.
67. See Daniel Pecaut, *Politica y Sindicalismo en Colombia* (Bogotá: La Carreta, 1973), p. 89.
68. For a broader discussion of this, see Miguel Urrutia, *The Development of the Colombian Labor Movement* (New Haven, CT: Yale University Press, 1969), p. 76.
69. Ibid., p. 16.
70. See Safford and Palacios, *Colombia,* p. 281.
71. See Defensoria del Pueblo—Colombia, *Antioquia: Nordeste,* p. 13.
72. See, for example, Urrutia, *Development of the Colombian Labor Movement,* p. 70.
73. See Pecaut, *Politica y Sindicalismo en Colombia,* p. 98.
74. See González, *Pasado y Presente del Sindicalismo Colombiano,* p. 88. See also Romero, *Paramilitaries y autodefensas,* pp. 170–174.
75. González, *Pasado y Presente del Sindicalismo Colombiano,* p. 137.
76. See Álvaro Delgado, *Politica y movimiento obrero, 1970–1983* (Bogotá: Ediciones CEIS, 1984), p. 113.
77. Ibid., p. 191.
78. See Edgar Caicedo, *Historia de las luchas sindicales en Colombia* (Bogotá: Ediciones CEIS, 1971), p. 231.
79. Jeremy Rayner, "Labor Rights in Colombia," Henning Center for International Labor

Relations, UC Berkeley, 2002, www.iir.berkeley.edu/henningcenter/gateway/colombia.html, viewed 3 May 2003.

80. Gobierno de Colombia, DANE, "Población Económicamente Activa y Número de Sindicalizados," pamphlet, Censo ENS, 2002.
81. Interview by author with Fanny Uribe, Oficial, Conferencia Episcopal de Colombia, Seccion de Movilidad Humana, Bogotá, 17 February 2004.
82. This figure is from Compania Colombiana de Datos and is contained in Oquist's *Violence, Conflict and Politics in Colombia,* p. 277. This source provides data based on the departmental origin of displaced persons, and some of the data include Antioquia, 116,500; Bogotá, 31,200 (city rather than department); Boyaca, 123,000; Cundinamaraca, 265,700; Huila, 112,000; Meta, 16,800; Norte de Santander, 174,400; Santander, 290,500; Tolima, 224,700; and Valle, 368,900.
83. See Richani, *Systems of Violence,* p. 66.
84. See Nora Segura Escobar, "Colombia: A New Century, an Old War, and More Internal Displacement," *International Journal of Politics, Culture and Society,* vol. 14, #1, 2000, pp. 111–112.
85. Codhes, *Un Pais Que Huye,* p. 148.
86. From 1985 to 1994, Codhes collected data along with the Catholic Church. Codhes began collecting its own data from 1994.
87. British Broadcasting Corporation News, 5 February 2004.
88. Interview by author with Harvey Suárez, Director, Codhes, Bogotá, 7 July 2003.
89. Codhes, *Un Pais Que Huye,* p. 157.
90. Ibid., p. 249.
91. Amnesty International, "Colombia: Just What Do We Have to Do to Stay Alive?" 1997, http://web.amnesty.org/library/print/ENGAMR230481997, viewed 3 October 2003.
92. Interview by author with Juan Pedro Schaerer, Jefe de la Delegacion, Comite Internacional de la Cruza Roja, Bogotá, 19 February 2004.
93. United Nations Educational, Scientific, and Cultural Organization (UNESCO), "Displacement, Development, and Modernity in the Colombian Pacific," pamphlet (London: UNESCO, 2003), pp. 158–159.

Chapter 3

1. By 2004, $2,568,280,000 had been devoted to military assistance, and about $639,460,000 to social and economic programs.
2. U.S. Department of State, "International Narcotics and Law Enforcement—FY 2006 Budget Justification," April 2005, http://state.gov.p/inl/rls/rpt/cbj/fy2006/511139.htm, viewed 3 May 2006; "Colombia," U.S. Department of Defense, Defense Security Cooperation Agency, 2001, www.dsca.mil/prgrams/Comptroller/2001_FACTS/, viewed 6 October 2005. The Center for International Policy provides up-to-date information on U.S. aid to Colombia on its excellent Web site. See Center for International Policy, Washington, DC, 2004, www.ciponline.org/colombia/aidtable/htm, viewed 6 October 2005; "Colombia Program—Military and Police Assistance Programs," 5 April 2006, viewed 3 May 2006; and "Colombia Program," 26 September 2006, viewed 12 January 2007. U.S. Department of State, "FY 2007 Congressional Budget Justification for Foreign Operations—Colombia," 13 February 2006, www.state.gov/documents/organization/606656.pdf, viewed 12 January 2007; and U.S. Department of State, "International Narcotics and Law Enforcement: FY 2007 Budget Justification—Colombia," www.state.gov/p/inl/rls/rpt/cbj/fy2007/71873.htm, viewed 12 January 2007.

3. Curt Tarnoff and Larry Nowels, Congressional Research Service, "Foreign Aid: An Introductory Overview of US Programs and Policy," 15 April 2004, p. 13, www.usinfo.state.gov/usa/infousa/trade/files/98-916.pdf, viewed 22 November 2005.
4. See, for example, David Rapoport, "Fear and Trembling: Terrorism in Three Religious Traditions," *The American Political Science Review,* vol. 78, #1, 1984, pp. 658–677.
5. Terror has also been used by numerous states, especially in South America. Dirty Wars in the Southern Cone, Central America, and Colombia are key examples.
6. For an overview of the concept of terror, see, for example, Alexander George, ed., *Western State Terrorism* (New York: Routledge, 1991); Martha Crenshaw, ed., *Terrorism in Context* (University Park: Pennsylvania State University Press, 1995); Tore Bjorgo, ed., *Terror from the Extreme Right* (London: Frank Cass, 1995); and Bruce Hoffman, *Inside Terrorism* (New York: Columbia University Press, 1998).
7. See GlobalSecurity.org, "Plan Colombia," 8 November 2004, www.globalsecurity.org/military/ops/colombia.htm, viewed 2 January 2005.
8. U.S. Department of State, "President Bush, President Pastrana Discuss Trade, Terrorism, George W. Bush, President, Remarks with President of Colombia at the White House," 18 April 2002, www.state.gov/p/wha/rls/rm/9542.htm, viewed 18 March 2003.
9. See GlobalSecurity.org, "Colombia," www.globalsecurity.org/military/ops/colombia.htm, viewed 16 August 2005.
10. For figures on training, see U.S. Department of State, "Foreign Military Training: Joint Report to Congress, Fiscal Year 2003–2004," June 2004, www.state/gov/t/pm/rls/rpt/fmtrpt/2004/34221.htm, viewed 3 October 2005. The figure provided includes an estimate of 4,258 soldiers in 2004. See also Adam Isacson, "Optimism, Pessimism, and Terrorism: The United States and Colombia in 2003," *Brown Journal of World Affairs,* vol. 10, #2, Winter 2004, p. 246.
11. See U.S. Defense Institute of Security Assistance Management, "Counterterrorism Fellowship Program," 2005, www.disam.dsca.mil/itm/Programs/CTF/@CTF_Program.htm, viewed 23 August 2005; and U.S. Department of State, "Foreign Military Training: Joint Report to Congress, Fiscal Year 2003–2004."
12. *Janes Intelligence Review,* 1 March 2004, www.janes2.com; and *New York Times,* 1 February 2005.
13. Comisión Internacional de Juristas, "Colombia: Políticos del Gobierno Socavan el Estado de Derecho y Consolidan la Impunidad," mimeograph, Bogotá, 21 September 2005, p. 30.
14. There were 64 fronts by mid-2005. Other organizational features of the FARC are blocks of up to 2,000 soldiers, as well as reserves in urban terrain.
15. Estimate is by the Washington, DC, NGO, International Campaign to Ban Landmines. See *New York Times,* 24 October 2004. Death rate from land mines is data from 2003, as reported in *El Tiempo,* 6 March 2004.
16. See James Rochlin, *Vanguard Revolutionaries in Latin America: Colombia, Peru and Mexico* (Boulder, CO: Lynne Rienner, 2003), chapters 2 and 3.
17. See U.S. Department of State, "Congressional Budget Justification for Foreign Operations, Fiscal Year 2005," 2004, www.state.gov/m/rm/rls/cbi, viewed 27 December 2005.
18. See *El Tiempo,* 10 December 2006.
19. See: U.S. Department of State, "A Report to Congress on United States Policy towards Colombia and Other Related Issues," 3 February 2003, www.state.gov/p/what/rls/rpt/17140.htm, viewed 3 July 2005; and U.S. Government Accountability Office (GAO), *Drug Control: Aviation Program Safety Concerns in Colombia Are Being Addressed, but State's Planning and Budgeting Process Can Be Improved* (Washington, DC: GAO, July 2004), p. 17.

20. República de Colombia, Ministerio de Defensa Nacional, "Descripción del apoyo de Estados Unidos al Plan Colombia," pamphlet, 2001.
21. U.S. Department of State, "Congressional Budget Justification for Foreign Operations, Fiscal Year 2005."
22. Ibid.; U.S. Department of State, "A Report to Congress on United States Policy towards Colombia and Other Related Issues."
23. See *El Tiempo,* 5 February 2006.
24. The process was obvious to me during my research in Colombia. It has been observed by many others, including Dr. Eduardo Pizarro Leóngomez, in an interview by author, Bogotá, 6 August 2004; and Adam Isacson, who has noted the trend in Arauca since 2001 and in western Cundinamarca since 2003, in addition to Medellín and Barrancabermeja. See Isacson, "Optimism, Pessimism, and Terrorism," pp. 250–251.
25. See *Los Angeles Times,* 25 March 2007.
26. See U.S. Department of State, "State Department Overview of Terrorist Threat in the Americas," 29 April 2004, http://usinfo.state.gov/is/Archive/2004/Apr/29-69799.html, viewed 23 May 2004.
27. Paul Virilio, *Speed and Politics* (New York: Semiotext(e), 1977), p. 18.
28. Interview by author with Andrés Dávila Ladron de Guvara, Director de Justicia y Seguridad, Gobernacion, Gobierno de Colombia, Bogotá, 16 February 2004.
29. Interview by author with Carlos Lobo, Director, Center of Information, Defense Systems Limited Colombia (DSC), Bogotá, 8 July 2003.
30. This is according to the Colombian army general Enrique Mora, *El Tiempo,* 8 November 2003.
31. Center for International Policy, "US Support for Plan Colombia," 2 September 2003, www.ciponline.org/facts/coaid.htm, viewed 13 October 2003.
32. *El Tiempo,* 18 June 2003.
33. *New York Times,* 31 December 2000.
34. U.S. Government, Bureau of International Narcotics and Law Enforcement Affairs, "Resource Summary," February 2005, www.whitehousedrugpolicy.gov/publications/policy/06budget/state.pdf, viewed 2 June 2005.
35. Pacific News Service, 10 July 2003.
36. *New York Times,* 25 July 2004.
37. See *Los Angeles Times,* 26 August 2003.
38. Interview by author with Rafael Nieto, Military Analyst, Bogotá, 8 July 2003.
39. See Ingrid Vaicius and Adam Isacson, "War on Drugs Meets War on Terror," pamphlet, Washington Office on Latin America, February 2003.
40. See, for example, *El Tiempo,* 16 August 2003.
41. *El Tiempo,* 29 June 2003.
42. Interview by author with Armando Borerro, Colombian Security Expert, Bogotá, 4 July 2003.
43. Interview by author with Rafael Nieto.
44. The Colombian government in March 2006 alleged FARC infiltration of 15 urban universities throughout the country. See *El Tiempo,* 22 March 2006.
45. See *New York Times,* 27 November 2004.
46. Interview by author with Carlos Lobo.
47. See *New York Times,* 5 January 2004.
48. Interview by author with Adam Isacson, Colombian Policy Analyst, Center for International Policy, Washington, DC, 29 July 2004.
49. Interview by author with Alfredo Rangel, Director, Seguridad y Democracia, Bogotá, 8

July 2003; and Francisco Leal Buitrago, Profesor, Universidad de los Andes, Bogotá, 10 July 2003.

50. Interview by author with Armando Borerro.
51. See *El Tiempo,* 16 February 2004.
52. LatinamericaPress.org, www.lapress.org, 5 February 2004.
53. FARC, Secretariado del Estado Mayor Central de las FARC-EP, "Saludo de las FARC," January 2006, www.farc-ep.ch/novedades/comunicados/sec2006/enero_01b.phb, viewed 3 March 2006.
54. Roman Ortiz, "Insurgent Strategies in the Post-Cold War: The Case of the Revolutionary Armed Forces of Colombia," *Studies in Conflict and Terrorism,* vol. 25, 2002, p. 141.
55. See *El Tiempo,* 25 April 2006.
56. As reported in *New York Times,* 28 October 2005.
57. As reported by ABC News, "Colombia," 18 December 2005, www.abc.net/au/news/newsitems/2005_12/S15335.htm, viewed 4 January 2006.
58. See *Los Angeles Times,* 11 March 2007.
59. Interview by author with Juan Bernardo Rosada Duque, Coordinador, Area de Derechos Humanos y Laborales, Escuela Nacional Sindical, Medellín, 12 August 2004.
60. Juan Bernardo Rosado Duque, Escuela Nacional Sindical, "Informe sobre la violación a los derechos humanos de los y las sindicalistas colombianos en el año 2005," mimeograph, Medellín, April 2006, pp. 11–12; also available on the group's Web site, www.ens.org.co.
61. Interview by author with Juan Ramos Rios, Secretario General, Unión Sindical Obrera de la Industria del Petroleo (USO), Bogotá, 4 August 2004; Julio Roberto Gómez, Presidente, Confederación General de Trabajadores Democráticos (CGTD), Bogotá, 4 August 2004; Ramón Támara Rivera, Director, Departamento de Relaciones Internacionales, Central Unitaria de Trabajadores de Colombia (CUT), Bogotá, 6 August 2004; Jorge Eliécer Ramírez, Director, Relaciones Publica, USO Nacional, Barrancabermeja, 9 August 2004; and Álvaro Torrado, Miembro de Directorate, USO Nacional, Barrancabermeja, 9 August 2004.
62. Ibid.
63. Interview by author with Eduardo Ortegon and Jaz Mile, La Comisión de Justica y Paz, Diocesis de Barrancabermeja, Barrancabermeja, 9 August 2004; Max Hill, Diocesis de Medellín, Medellín, 12 August 2004; and Harvey Suárez, Director, Codhes, Bogotá, 16 February 2004.
64. *Caracol,* 31 January 2006.
65. See *El Tiempo,* 2 November 2005 and 2 December 2005.
66. Statistics from United Nations Development Program (UNDP), "Colombia 2005," www.hdr.undp.org/statistics/data/countries.cfm?c=COL, viewed 3 May 2006; statistics reported by the UNDP in *El Tiempo,* 7 September 2005; and Rochlin, *Vanguard Revolutionaries in Latin America,* pp. 14 and 99.
67. See *El Tiempo,* 12 April 2004.
68. See, for example, *El Tiempo,* 16 February 2005.
69. Gabriel García Márquez, *One Hundred Years of Solitude* (New York: Avon, 1971), p. 31.

Chapter 4

1. See, for example, P. W. Singer, *Corporate Warriors: The Rise of the Privatized Military Industry* (Ithaca, NY: Cornell University Press, 2003), p. 220.

2. See James Rochlin, *Vanguard Revolutionaries in Latin America: Peru, Colombia and Mexico* (Boulder, CO: Lynne Rienner, 2003), pp. 99–100.
3. Center for Defense Information, transcript of ITV broadcast (Britain), "World in Action—BP's Secret Soldiers," broadcast 8 October 1997, www.cdi.org/armstrade database/CONTROL/Small_Arms/mercenaries/BP's_Secret_Soldiers.txt, viewed 2 January 2002. For an in-depth discussion of British Petroleum (BP)'s relationship with DSC, and surrounding controversies, see www.carbonweb.org/documents/chapter11 .pdf, viewed 2 February 2003.
4. See Singer, *Corporate Warriors,* p. 221.
5. Human Rights Watch, "Colombia: Human Rights Concerns Raised by the Security Arrangements of Transnational Oil Companies," April 1998, www.hwr.org/advocacy/ corporations/colombia/Oilpat.htm, viewed 2 February 2003.
6. *Guardian* (Britain), "BP in Colombia," 15 August 1998.
7. See Nazih Richani, *Systems of Violence: The Political Economy of War and Peace in Colombia* (Albany: SUNY, 2002), p. 116.
8. The Center for Public Integrity, "Colombia," 20 October 2004, www.store.public integrity.org/report.aspx?aid=262$sid=100, viewed 1 December 2004.
9. *El Tiempo,* 9 December 2000.
10. Stratfor.com, "US Presence in Colombia," www.stratfor.com/hotspot_Colombia/ presence.htm, viewed 5 October and 12 November 2003.
11. *El Tiempo,* 18 June 2003; and *New York Times,* 12 September 2003.
12. U.S. Department of State, "Report on Certain Counternarcotics Activities in Colombia," April 2003, www.ciponline.org/colombia/03041401.htm, viewed 2 November 2003.
13. J. McDermott, "Uribe Gains the Upper Hand in Colombia's Guerrilla War," *Janes Intelligence Review,* 1 December 2003, www.janes.com.
14. Center for Public Integrity, "Colombia," www.store.publicintegrity.org/reports.aspx ?aid=262$sid=100, viewed 20 October 2004 and 23 January 2005.
15. Interview by author with Jean Pineault, Security Manager, Nexen Corporation, Bogotá, 8 July 2002.
16. Interview by author with Carlos Lobo, Director, Center of Information, DSC, Bogotá, 8 July 2003.
17. Interview with Carlos Lobo, Bogotá, 11 July 2002.
18. The director of DSC, however, refused to grant an interview, and did not reply to questions by e-mail, which would have allowed him to respond to allegations regarding his company's relations to paramilitary organizations or the provision of intelligence to the Colombian military and paramilitary organizations regarding individuals who opposed petroleum projects of transnational corporations in Colombia.
19. Regarding the issue of accountability in Colombia, there was some fuss generated by the Airscan scandal, but little overall interest. Interview by author with Armando Borerro, Security Expert, Bogotá, 4 July 2003; Alfredo Rangel, Director, Fundación Seguridad y Democracia, Bogotá, 9 July 2003; and Dr. Francisco Leal Buitrago, Profesor, Ciencia Politica, Universidad de los Andes, Bogotá, 10 July 2003.
20. Interview with Adam Isacson, Colombian Policy Analyst, Center for International Policy, Washington, DC, 29 July 2004.
21. FARC, FARC-EP, "Empresas criminals avalades por Seguridad Democrática," 11 November 2004, www.farcep.org/novedades/coyuntura/paramilitarismo/noviembre_11 _2004.php, viewed 22 March 2005.
22. U.S. Department of State, "Fiscal Year 2007 Congressional Budget Justification—Western Hemisphere, Colombia," 13 February 2006, p. 14, www.state.gov/documents/ organization/60656.pdf, viewed 29 May 2006.

23. Fernando Vallejo, *Our Lady of the Assassins* (London: Serpent's Tale, 2001), p. 73.
24. An excellent work on this topic is Russell Crandall, *Driven by Drugs: United States Policy toward Colombia* (Boulder, CO: Lynne Rienner, 2002).
25. See *El Tiempo,* 2 May 2005; and Jeremy McDermott, "Farc and the Paramilitaries Take Over Colombia's Drug Trade," *Janes Intelligence Review,* 1 July 2004.
26. See *New York Times,* 25 April 2007.
27. As reported in *El Tiempo,* 30 November 2005.
28. U.S. Department of State, "Colombian Narco-Terrorist Extradited to the United States," 31 December 2004, www.usinfo.state.gov/wh/archive/2005/Jan/03-115430.html, viewed 13 June 2005.
29. The United States charged 50 FARC members, including Marulanda, Raul Reyes, Mono Jojoy, and Alfonso Cano, and has requested their extradition to the country.
30. As reported in the *New York Times,* 5 March 2006.
31. See *New York Times,* 2 August 2004.
32. For example, some have observed cooperation between the FARC and AUC in 2004 regarding drug flows through the Magdalena Media and Nariño. See McDermott, "FARC and Paramilitaries Take Over Colombia's Drug Trade."
33. *New York Times,* 17 June 2004.
34. *El Tiempo,* 15 April 2005.
35. See *El Tiempo,* 30 December 2006.
36. U.S. Government, Office of National Drug Control Policy, "Technical Report for Price and Purity of Illicit Drugs," Appendix H, 2004, www.whitehousedrugpolicy.gov/publications/price%5Ftech%5Frpt/apph.pdf, viewed 10 June 2005. A careful and excellent analysis of this information is offered by Coletta Youngers and Eileen Rosen, eds., *Drugs and Democracy in Latin America: The Impact of US Policy* (Boulder, CO: Lynne Rienner, 2004). See also, Youngers and Rosen, *Drugs and Democracy in Latin America,* A WOLA special report (Washington, DC: Washington Office on Latin America, November 2004).
37. UK Parliament, "Drugs," 15 February 2006, www.publications.parliament.uk/pa/cm200506/cmhansard/cm060215/text/60215w27.htm, viewed 28 May 2006.
38. As reported in *El Tiempo,* 4 August 2004.
39. On the issue of glysophate and its toxic effects, see *Un Periódico,* Universidad Nacional de Colombia, #75, 22 May 2005.
40. Al Geddicks, "Resource Wars against Native Peoples in Colombia," *Capitalism, Nature, Socialism,* vol. 14, #2, June 2003, p. 101.
41. For a broader discussion of this, see Plinio Apuleyo Mendoza, "Las FARC: Un Enclave Terrorista en Colombia," in *Terrorismo y Seguridad,* ed. Reinaldo Botero (Bogotá: Planeta, 2003), pp. 195–210.
42. See, for example, Steven B. Karch, *A Brief History of Cocaine* (New York: CRC Press, 1998); and Richard Davenport-Hines, *The Pursuit of Oblivion: A Global History of Narcotics, 1500–2000* (London: Weidenfeld and Nicolson, 2001).
43. As quoted in National Hydrocarbon Agency, Government of Colombia, "Colombian Oil and Gas Investment Conference, 2005," May 2004, www.gematours.com/hidrocarburos/en/paper/htm, viewed 12 September 2005.
44. *Caracol Noticias,* 7 February 2005.
45. President Uribe admitted that weak intelligence and surveillance were among the military errors that permitted the attack. See *El Tiempo,* 27 June 2005.
46. International Monetary Fund (IMF), "Colombia," Country Report, #05/162, May 2005, www.imf.org/external/pubs/ft/scr/2005/cr05162.pdf, viewed 14 June 2005.
47. For a good general discussion of the Colombian oil industry, see Garry Leech, "Plan Petroleum in Putumayo," *Nacla Report on the Americas,* vol. 37, #6, May 2004, pp. 8–11.

48. Quoted in the *New York Times,* 12 January 2005.
49. Álvaro Mutis, *The Adventures and Misadventures of Maqroll* (New York: New York Review Books, 2002), p. 407.
50. For a broader discussion of this, see Jeff Browitt, "Capital Punishment: The Fragmentation of Colombia and the Crisis of the Nation-State," *Third World Quarterly,* vol. 22, #6, 2001, p. 1063.
51. Interview by author with Juan Bernardo Rosada Duque, Coordinador, Area de Derechos Humanos y Laborales, Escuela Nacional Sindical, Medellín, 12 August 2004.
52. Escuela Nacional Sindical, "Cae Dramáticamente la Creación de Sindicatos en Colombia Durante el Gobierno Uribe," mimeograph, Medellín, 12 May 2006, p. 1.
53. Interview by author with Julio Roberto Gómez, Presidente, CGTD, Bogotá, 5 August 2004.
54. Interview by author with Juan Bernardo Rosada Duque.
55. Interview by author with Norman Shipull, Director, Oficina Internacional del Trabajo, U.S. Department of Labor, Bogotá, 5 August 2004.
56. Interview by author with Juan Ramon Rios, Presidente, USO, Bogotá, 14 July 2006.
57. *El Tiempo,* 20 March 2007.
58. Interview by author with Tarsicio Mora Fodoy, Fiscal de la CUT, Bogotá, 12 July 2006.
59. For a broader discussion of this, see Meredith Aby, "Colombia's Unions: Under Attack and Fighting Back," 22 August 2004, www.fightbacknew.org, viewed 6 September 2004.
60. See *El Tiempo,* 15 March 2004.
61. Interview by author with Julio Roberto Gómez.
62. Interview by author with Ramón Támara Rivera, Director, Departamento de Relaciones Internacionales, CUT, Bogotá, 6 August 2004.
63. Interview by author with Juan Ramon Rios, Secretario General, USO, Bogotá, 4 August 2004.
64. Interview by author with Jorge Elicier Ramírez, Director, Relaciones Publicas, USO Nacional, Barrancabermeja, 9 August 2004; and Álvaro Torrado, Miembro de Directorate, USO Barrancabermeja, Barrancabermeja, 9 August 2004. For additional information regarding the rise of the AUC in Arauca after 2001, see Conferencia Episcopal de Colombia, Boletin Especial, *Estudio Caso 5: Arauca* (Bogotá: Conferencia Episcopal de Colombia, 2003), p. 7.
65. Interview by author with Juan Ramon Rios, 4 August 2004.
66. As reported in *El Tiempo,* 28 May 2004.
67. Interview by author with Hector Yesid Vaca Cespedes, Presidente, and Gabriel Alvis, Expresidente, Junta Nacional de la USO, Bogotá, 24 May 2005.
68. See *El Tiempo,* 20 November 2004.
69. See, for example, ELN, "Conclusiones del sexto pleno de la direccion nacional del ELN," October 2004, www.eln.voces.com/Correa_del_Magdalena/Revista%Unidad%20VI.htm; ELN, "2004—Año de Combate Social," *Revista,* #25, 23 December 2004, www.eln.voces.com/Insurreccion/Nacional.html; and FARC, "Plan Colombia: Brazo militar del ALCA, sin resultados," 16 August 2003, www.farcep.org/novedades/coyuntura/plan_colombia/agosto162003.php, all viewed 10 January 2005.
70. See *El Tiempo,* 7 March 2007.
71. Gabriel García Márquez, *Living to Tell the Tale* (New York: Alfred A. Knopf, 2003), p. 355.
72. This estimate of the displaced is from Codhes, "Comportamiento de Desplazamiento 1985–2005," April 2006, www.codhes.org.co/cifra/GraficoTendencias1985_2005.jpg, viewed 10 May 2006.
73. This has occurred especially since the power grab of the 1990s and has continued into the new millennium. Some 3 million hectares of land are estimated to have been aban-

doned by the displaced population between 1996 and 1999. See Codhes, *Un Pais Que Huye: Desplazamiento y violencia* (Bogotá: Codhes, 2003), p. 181.

74. Interview by author with Harvey Suárez, Director, Codhes, Bogotá, 16 February 2004.
75. See Oficina del Alto Comisionado de las Naciones Unidas para los Refugiados (ACNUR), *La Población Desplazada en Bogotá* (Bogotá: ACNUR, 2003), p. 33.
76. Codhes, *Un Pais Que Huye*, p. 181.
77. Interview by author with Harvey Suárez, Bogotá, 7 July 2003.
78. ACNUR, *La Población Desplazada en Bogotá*, p. 43.
79. Interview by author with Fanny Uribe, Oficial, Conferencia Episcopal de Colombia, Seccion de Movilidad Humana, Bogotá, 17 February 2004.
80. Interview by author with Ana Ferrero, Oficial de Protección, ACNUR, Bogotá, 8 July 2003.
81. Interview by author with Harvey Suárez, Bogotá, 16 February 2004. This was also the case with the displaced families I interviewed, who suggested that it was the "junta," or the collection of armed groups fighting among themselves, that propelled their expulsion from the Bucaramanga region of Santander. Interview by author with displaced families, arranged by Conferencia Episcopal de Colombia, Bogotá, 20 February 2004.
82. Codhes, *Un Pais Que Huye*, p. 298.
83. Interview by author with Fernando Calado Bryce, Oficial de Programa, Organización para las Migraciones (OIM), Bogotá, 17 February 2004.
84. Codhes, *Un Pais Que Huye*, p. 415. See also Conferencia Episcopal de Colombia, *Estudio Caso 5*.
85. International Red Cross, *Estadisticas: Asistencia Humanitaria de Emergencia Colombia* (Geneva: International Red Cross, 2003), p. 2.
86. Arauca Guahibo Indian leader Dario Tuliva, quoted in Bill Weinberg, "Arauca: The Attack in Colombia's Oil Zone," *ZNET*, 8 September 2003, www.smag/org/content/showarticle.cfm?SectionalID=9$ItemID=4149, viewed 12 February 2004.
87. See Al Geddicks, "Resource Wars against Native Peoples in Colombia," p. 88.
88. See Defensoria del Pueblo—Colombia, Secretariado Nacional de Pastoral Social Caritas Colombiana, *Situacion de los DDHH y del Derecho Humanitario en la Sierra Nevada de Santa Marta* (Bogotá: Defensoria del Pueblo, 2003). For example, between 1998 and 2002, 44 indigenous disappeared, 166 were executed, 92 tortured, and 52 kidnapped.
89. Harvey Suárez, Codhes, "El Desplazamiento Forzada Indigena en Colombia," mimeograph, Bogotá, 2001.
90. ACNUR, *La Población Desplazada en Bogotá*, p. 69.
91. Codhes, "El Aceite Amargo de la Palma Aceitera," 24 June 2005, www.codhes.org.co/dbreves.php?breve=627, viewed 23 August 2005. See also, Conferencia Episcopal de Colombia, "Informa Sobre Desplazamiento Forzado en Colombia," pamphlet, Bogotá, July 2002.
92. Interview by author with Jattan Mazzot, Director, Asociación de AfroColombianos Desplazados, Bogotá, 7 July 2003.
93. This point is commonly made among NGOs who work with Colombia's displaced population. See, for example, Codhes, *Un Pais Que Huye*, p. 146.
94. Secretariado del Estado Central de las FARC-EP, "Comunicado de las FARC," 9 March 2003, www.farc-ep.ch/comunicados/, viewed 24 May 2003.
95. The IRA members were originally released by Colombian authorities in June 2004 but told to remain in the country until a subsequent legal hearing was conducted in December 2004, which linked the IRA members to the training of FARC guerrillas. By that time, they apparently fled the country.

96. See Human Rights Watch, "Colombia: Farc Steps Up Attacks before Election," 6 March 2006, www.hrw.org/english/docs/2006/03/06/colomb12761.txt.htm, viewed 24 March 2006; and *El Tiempo*, "Conferencia Episcopal Colombia rechaza ataques de las guerrilla contra civiles," 4 July 2005.
97. See, for example, FARC-EP, Pleno del Estado Mayor, "Declaración Politica—la patria esta amenazada," March 2000, www.six.swix.ch/farcep/Pleno/declaracion.html, viewed 30 September 2001; and FARC-EP, Comisión Temática de las FARC-EP, "A la Política Agraria de Mentiras de la Oligarquia, oponemos una efectiva polítca agraria revolucionaria," March 2000, www.six.swix.ch/farep/audiencias/reforma.html, viewed 24 April 2003.
98. See, for example, FARC-EP, Secretariado del Mayor Central FARC-EP, "Comunicado," December 2004, www.farcep.org/novedades/comunicados/sec2004/deciembre4.php; and FARC-EP, "Plan Colombia y amanaza para todas," 13 August 2001, www.six.swix.ch/farcep/Articulos/plancolombia.html, both viewed 29 December 2004.
99. See: FARC-EP, "Empresas criminales avalades por Seguridad Democratica," 11 November 2004, www.farcep/org/novedades/coyuntura/paramilitarismo/noviembre_11_2004.php; and Juan Leonel, FARC member, "Paramilitarismo y multinacionales," 21 May 2004, www.farcep.org/novedades/coyuntura/editorial/mayo_21_2004.php, both viewed 10 December 2004.
100. Raul Reyes, "Los pueblos son las victimas de Terrorismo del Estado de los Gobiernos Capitalistas," 18 February 2003, www.farcep.org/documentos/docs/victimas1802.htm, viewed 11 April 2004.
101. FARC, Comision Internacional de Las FARC-EP, "Socialismo, Experiencias en el Cambio e Integración," 18 May 2006, www.farcep.org/?node=2,2059,1, viewed 29 May 2006.
102. Interview by author with Eduardo Pizarro Leóngomez, Presidente, Comisión Nacional de Reparación y Reconciliación, Bogotá, 14 July 2006.
103. See ELN, "Correo del Magdalena," no. 162, 15 January 2004, www.eln-voces.com/Insurreccion/Anteriores/CM/2004/CM161.htm; ELN, "El Sindicalismo Colombiano: En la Mira de la Democracia Paramilitar," 6 July 2004, www.eln-voces.com/Insurreccion/Nacional3.html; and ELN, "Pronunciamiento," 18 December 2004, www.eln-voces.org/Correa_del_Magdalena/Pronunciamiento-Trinidad-18-12-04.htm, all viewed 5 January 2005.
104. See, for example, ELN, "Coyuntura Nacional: Arauca; Prioridad del Plan Colombia," 22 December 2003, www.eln-voces.com/Insurreccion/Anteriores/CM?2003?CM161.htm, viewed 22 June 2004.
105. As reported in *El Tiempo,* 7 May 2004.
106. ELN, "Comunicado: Contra la Guerra de Uribe Cumplimos con la Paz," 22 December 2003, www.eln-voces.com/Insurreccion/Anteriores/CM/2003/CM161.htm, viewed 16 February 2004.
107. Interview by author with Armando Borerro.
108. ELN, "Conclusiones del sexto pleno de la direccion nacional del ELN," October 2004, www.eln-voces.org/Correa_del_Magdalena/Revista%Unidad%20VI.htm, viewed 14 March 2005.
109. ELN, "Con Impunidad, No Hay Paz," 26 May 2006, www.eln-voces.com/Insurreccion/Editorial.html, viewed 29 May 2006.
110. Vallejo, *Our Lady of the Assassins,* p. 16.
111. AUC, "Discurso del Jefe del Estado Mayor de las AUC Comandante Salvatore Mancuso, en el acto de Instalacion Oficial del Proceso de Negociacion entre el Gobierno Nacional y las AUC," 1 July 2004, www.colombialibre.org/ver_imp.php?Varid=6425, viewed 2 September 2004.

112. See *New York Times,* 4 July 2005.
113. Interview with Eduardo Pizarro Leóngomez, Profesor, Instituto de Estudios Políticos y Relaciones Internacionales, Universidad Nacional de Colombia, Bogotá, 6 August 2004.
114. Eduardo Pizarro Leóngomez, *Una democracia asediada—balances y perspectivas del conflicto armado en Colombia* (Bogotá: Norma, 2004), p. 147. The author argues that the AUC has at times behaved as a terrorist organization.
115. See *New York Times,* 11 December 2004.
116. See Mauricio Romero, *Paramilitares y autodefensas, 1982–2003* (Bogotá: IEPRI, 2003), p. 17.
117. See *El Tiempo,* 2 July 2005.
118. AUC, Estado Mayor Negociador, Salvatore Mancuso, "Nuestra Propuesta de Incorporación a las Fuerzas Armadas," 26 October 2004, www.colombialibre.org/ver_imp.php?Varid=8398, viewed 10 November 2004.
119. As reported in *El Tiempo,* 25 April 2006.
120. *El Tiempo,* 30 September 2005.
121. *Caracol News,* 24 June 2005.
122. *New York Times,* editorial, "Colombia's Capitulation," 4 July 2005.
123. Comisión Internacional de Juristas, "Colombia: Políticos del Gobierno Socavan el Estado de Derecho y Consolidan la Impunidad," mimeograph, Bogotá, 21 September 2005, pp. 77 and 23.
124. See *New York Times,* 19 March 2007.
125. See *New York Times,* 10 March 2007.
126. As reported in *El Tiempo,* 1 February 2007.
127. *Caracol,* 12 February 2007.
128. See *New York Times,* 28 May 2006. During the first two years of the Uribe administration there was a decline in the number of homicides (16 percent), massacre victims (52 percent), and massacres (49 percent), as well as assassination of the indigenous (20 percent), assassination of teachers (18 percent), assassination of journalists (41 percent), assassination of labor union members (72 percent), attacks by armed groups against towns (82 percent), attacks against communication towers (35 percent), attacks against energy towers (34 percent), attacks against oil pipelines (30 percent), and bombings of water lines (50 percent). *El Tiempo,* 20 December 2004.
129. See United Nations, Departamento Población, Equipo Coordinardor: Informe Nacional de Desarrollo Humano, *Andes 2020 y el Conflicto, callejón con salida: Una Mirada Comparativa* (Bogotá: UN Departamento Población, 2004).
130. Interview by author with Ana Teresa Bernal, Presidencia Colegiada, Redepaz, Bogotá, 6 August 2004.

Chapter 5

1. See *La Jornada,* 23 August 2005.
2. Quoted in *La Jornada,* 9 February 2007.
3. See *El Sur,* 28 July 2005. The state of Veracruz joins the other three as hosting the worst inequity in Mexico, according to the UNDP. Further, one-fourth of all extreme poverty in Mexico occurs in Chiapas, Guerrero, and Oaxaca. See *La Jornada,* 7 December 2003.
4. See *El Sur,* 25 June 2003.
5. This idea is expressed in two Mexican classics. See Carlos Fuentes, *The Buried Mirror: Reflections on Spain and the New World* (New York: Mariner, 1999), pp. 40, 100. See also Octavio Paz, *The Labyrinth of Solitude* (New York: Grove Press, 1985), p. 15.

6. See, for example, Jorge Hernández-Díaz, *Reclamos de la Identidad: La Formación de las Organizaciones Indígenas en Oaxaca* (Oaxaca: Universidad Autónoma Benito Juárez de Oaxaca, 2001), pp. 20–23.
7. See Paz, *Labyrinth of Solitude,* quote from p. 56. See also pp. 9, 20.
8. Ibid., p. 155.
9. See Juan Rulfo, *Pedro Páramo* (New York: Grove, 1994), p. 5. The original Spanish publication is from 1955.
10. For a broader discussion of this, see Armando Bartra, *Guerrero Bronco: Campesinos, ciudadanos y guerrilleros en la Costa Grande* (New Haven, CT: Yale University Press, 1996), p. 21.
11. See Machiavelli's *The Prince* (New York: Mentor, 1980), p. 92.
12. Paz, *Labyrinth of Solitude,* p. 144.
13. See Fuentes, *Buried Mirror,* p. 300.
14. For a larger discussion of this, see Courtney Jung, "The Politics of Indigenous Identity: Neoliberalism, Cultural Rights, and the Mexican Zapatistas," *Social Research,* vol. 70, #2, 2003, pp. 437–438.
15. For a broader discussion of this, see Sergio Aguayo Quezada, *La Charola: Una historia de los servicios de inteligencia en Mexico* (Mexico, DF: Grijalbo, 2001), p. 68.
16. For a broader discussion of this, see James Rochlin, *Redefining Mexican Security: Society, State and Region under Nafta* (Boulder, CO: Lynne Rienner, 1997).
17. For a broader discussion of this, see James Rochlin, *Vanguard Revolutionaries in Latin America: Peru, Colombia and Mexico* (Boulder, CO: Lynne Rienner, 2003), chapters 6 and 7.
18. See, for example, John Bailey and Jorge Chabat, "Introduction," in *Crimen Transnacional y Seguridad Pública,* ed. John Bailey and Jorge Chabat (Mexico, DF: Plaza y Jánes, 2003), pp. 16 and 28.
19. On the eve of a Latin American summit with the European Union, for example, Mexico's president Fox sparred with Venezuela's Hugo Chavez and Bolivia's Evo Morales. See *New York Times,* 11 May 2006.
20. See "China Values Co-op with Latin American Nations," www.chinaview.cn, viewed 31 March 2006.
21. See M. Ayhan Kose et al., "How Has NAFTA Affected the Mexican Economy?" (IMF working paper, #WP/04/59, April 2004), especially p. 29.
22. Juan Rulfo, "They Gave Us the Land," in *The Burning Plain and Other Stories* (Austin: University of Texas Press, 2003), p. 11.
23. See Armando Bartra, "Sur Profundo," in *Crónicas del sur: Utopías campesinas en Guerrero,* ed. Armando Bartra (Mexico, DF: Ediciones Era, 2000), p. 22.
24. For a broader historical discussion of *caciques* and political exclusion, see Jorge Rendón Alarcón, *Sociedad y conflicto, en el estado de Guerrero, 1911–1995* (Mexico, DF: Plaza y Valdes, 2003), pp. 21–35.
25. Procuraduría General de la República (Ofincina de Dr. Ignacio Carrillo Prieto), Fiscalía Especial para Movimientos Sociales y Políticos del Pasado, *Informe Histórico a la Sociedad Mexicana,* November 2006, p. 285.
26. Ibid., pp. 284–286.
27. Moisés de la Peña, *Guerrero económico* (Chilpancingo: Gobierno del estado de Guerrero, 1949), p. 456.
28. See Bartra, *Guerrero Bronco,* p. 86.
29. Ibid., p. 152.
30. *Informe Histórico a la Sociedad Mexicana,* p. 302.
31. For a good general discussion of Lucio Cabañas and his followers, see José Natividad

Rosales, *Quien es Lucio Cabañas* (Mexico, DF: Posada, 1974); and Carlos Bonilla Machorro, *Ejercicio de Guerrillero* (Mexico, DF: Gaceta, 1981).

32. Comité Central del Partido Democrático Popular Revolucionario y Comandancia General del Ejército Popular Revolucionario (PDPR CGEPR), "Un Poco Mas de Historia," September 2005, http://usarios.lycos.es/cedema/Mexico/EPR/epr(2005septiembre).html, viewed 7 October 2005.
33. See Simón Hipólito, *Guerrero, Amnistia y Represión* (Mexico, DF: Grijalbo, 1982), p. 31.
34. Luis Suárez, *Lucio Cabañas, el guerrillero sin esperanza* (Mexico, DF: Roca, 1978), p. 317.
35. Lucio Cabañas, "Así mi fui a la Sierra," in Suárez, *Lucio Cabañas,* p. 58.
36. Ibid., p. 57.
37. Cabañas, "Asamblea sobre el partido y la propaganda," in Suárez, *Lucio Cabañas,* p. 137.
38. The directive of El Partido de los Pobres (PDLP), beyond Cabañas, consisted of those who went under the names of "César," "Eduardo," "Luis," and "El Doc."
39. See Eleazar Campos Gómez, *Lucio Cabañas y el Partido de los Pobres* (Mexico, DF: Editorial Nuestra America, 1987), pp. 197 and 264. This book is composed of testimonials from members of the PDLP. The author is a Comandante with the PDLP.
40. Cabañas, "Imagen del Che Guevara," in Suárez, *Lucio Cabañas,* pp. 127–130.
41. Cabañas, "Así mi fui a la Sierra," p. 57.
42. Che Guevara, *Guerrilla Warfare* (Lincoln: University of Nebraska, 1998), p. 7.
43. See Gómez, *Lucio Cabañas y el Partido de los Pobres,* pp. 10–11.
44. Paulo Freire, *Pedagogy of the Oppressed* (New York: Continuum, 1982), p. 33.
45. See Rochlin, *Vanguard Revolutionaries in Latin America,* chapters 2 and 3.
46. Cabañas, "La justicia de Lucio," in Suárez, *Lucio Cabañas,* p. 175.
47. Machorro, *Ejercicio de Guerrillero,* p. 132.
48. *Informe Histórico a la Sociedad Mexicana,* p. 322.
49. See Cabañas, "Así mi fui a la Sierra," p. 69; and Cabañas, "Las tres primeras emboscadas," in Suárez, *Lucio Cabañas,* p. 118.
50. *Informe Histórico a la Sociedad Mexicana,* p. 328.
51. Quoted in Gómez, *Lucio Cabañas y el Partido de los Pobres,* p. 184.
52. Cabañas, "Informe de Lucio sobre la situación internacional, en la Segunda Asamblea General del Partido de los Pobres," in Suárez, *Lucio Cabañas,* p. 119.
53. *Informe Histórico a la Sociedad Mexicana,* pp. 499–503.
54. Ibid., p. 327.
55. Ibid., p. 317.
56. Ibid., p. 339. For example, at least 90 people were detained in El Quemado between 28 August and 5 September 1972.
57. Ibid., p. 336.
58. Ibid., p. 339.
59. Ibid., p. 427.
60. Ibid., p. 364.
61. Ibid., pp. 369–370.
62. See Quezada, *La Charola,* pp. 105–106.
63. Ibid., pp. 131–140.
64. Ibid., p. 140.
65. Quezada, *La Charola,* p. 202.
66. Bartra, *Guerrero Bronco,* p. 149.
67. PDPR CGEPR, "Un Poco Mas de Historia."
68. For a broader discussion of this, see Moisés Ochoa Campos, *Guerrero: Análisis de un estado problema* (Mexico, DF: Trillas, 1964).
69. See Rochlin, *Redefining Mexican Security,* pp. 24–25.

70. For a broader discussion of the collapse of agricultural support programs during this period, see Adán Aguirre Benítez, *Guerrero: Economia Campesina y Capitalismo, 1960–1987* (Chilpancingo: Universidad Autónoma de Guerrero, 1995), pp. 241–249.
71. PDPR CGEPR, "Un Poca Mas de Historia."
72. Ibid.
73. PDLP Comandantes Jose Luis Orbe Diego, Isidro Castro Fuentes, Adela Alvarez Rios, and Enrique Velazquez Fierrero, in Gómez (himself a Comandante of the PDLP), *Lucio Cabañas y el Partido de los Pobres,* pp. 9–10.
74. Ibid.
75. Ibid.
76. U.S. Department of Defense, Joint Staff, Washington, DC, 28 April 1992, "Mexican Defense Headquarters, Concern over Insurgents," documents released under Freedom of Information Act in August 1996 by *Proceso,* Kate Doyle, ed., "Rebellion in Chiapas and the Mexican Military," National Security Archive Electronic Briefing, Document #109, 20 January 1994.
77. Gisela Espinosa Damian and Miguel Meza Castillo, "Guerrero en Cifras," in Bartra, *Crónicas del sur,* pp. 89 and 98.
78. See Alejandro Martínez Carvajal, *Ejército Popular Revolucionario* (Mexico, DF: Editorial Sagitario, 1998), p. 8.

Chapter 6

1. Miguel Cervantes, *Don Quixote* (New York: Oxford University Press, 1999), p. 1112.
2. For an up-to-date and excellent overview of Mexican economic statistics, see Instituto Nacional Estadística Geografía e Informática (INEGI), "Estadísticas Económicas, Cuadro resumen," www.inegi.gob.mx.
3. Debt crisis is used here in the sense of an acute event.
4. See *La Jornada,* 4 May 2006.
5. See *La Jornada,* 13 February 2006, and *Miami Herald,* 1 February 2006, for quote by Alejandro Werner.
6. Dow Jones News, 16 February 2006.
7. *La Jornada,* 16 February 2006.
8. See James Rochlin, *Vanguard Revolutionaries in Latin America: Peru, Colombia and Mexico* (Boulder, CO: Lynne Rienner, 2003), especially chapters 6 and 7. See also, James Rochlin, *Redefining Mexican Security under Nafta: Society, State and Region* (Boulder, CO: Lynne Rienner, 1997).
9. Organization for Economic Co-operation and Development, "Economic Survey of Mexico 2005," 12 September 2005, www.oecd.org, viewed 2 November 2005.
10. INEGI, "Encuesta Nacional de Ingresos y Gastos de los Hogares," 2006, www.inegi.gob.mx, viewed 10 April 2006. For other information regarding ginis, see Luxemburg Income Study, 31 January 2006, www.lisproject.org/keyfiguers/ineqtable.htm; and *La Jornada,* 15 June 2005.
11. United Nations Economic Commission for Latin America and the Caribbean (ECLAC), *Notes,* #43, November 2005, pp. 1–6.
12. For a broader discussion of this, see Manuel Pastor, Jr., and Carol Wise, "The Lost Sexenio," *Latin American Politics and Society,* vol. 47, #4, 2005, pp. 144–145.
13. Maquiladoras are responsible for the manufactured portion of total Mexican merchandise exports, which rose from 43 percent in 1990 to 81 percent in 2003. Statistics from the UNDP, *Human Development Report 2005,* www.hdr.undp.org, viewed 6 April 2006.

14. INEGI, "Personal ocupado en la industria maquiladora de exportacion segun tipo de ocupacion," 2006, www.inegi.gob.mex. viewed 4 January 2007. See also Federal Reserve Bank of Dallas, "Maquiladora Industry Update," October 2005, www.dallasfed.org/data/data/maq_charts.pdf, viewed 15 January 2006.
15. See Gordon Hanson, "The Evolution of Mexico's Manufacturing Exports" (seminar papers, Economics Department, UC Davis, February 2006).
16. See *La Jornada,* 30 January 2006.
17. M. Ayan Kose et al., "How Has Nafta Affected the Mexican Economy?" (IMF working paper, #WP/04/59, April 2004), p. 29.
18. See *The Nation,* 2 February 2004, www.thenation.com, viewed 14 February 2006; and Bill Lambrecht, "Low Prices Force Mexicans from Fields," 30 October 2005, www.bilaterals.org/artcile.php3?id_artcile=3009, viewed 1 February 2006.
19. Steven Zahniser and William Coyle, "US Mexico Corn Trade during the Nafta Era," U.S. Department of Agriculture, FDS-04D-01, May 2004, p. 6.
20. See Gisela Espinosa Damian and Miguel Mexa Castillo, "Guerrero en Cifras," in *Crónicas del sur: Utopías campesinas en Guerrero,* ed. Armando Bartra (Mexico, DF: Ediciones Era, 2000), p. 80.
21. This comparison does not include revenue from narcotrafficking.
22. Interview by author with Álvaro López, Departamento de Historia, Universidad Autónoma de Guerrero, Chilpancingo, 20 May 2004.
23. L. Ortiz-Lozano et al., "Environmental Evaluation and Developmental Problems of the Mexican Coastal Zone," *Ocean and Coastal Management,* vol. 48, 2005, p. 173.
24. *Dallas Morning News,* 10 March 2003.
25. International Press Service, 31 May 2003.
26. *La Jornada,* 5 February 2006.
27. *El Sur,* 24 April 2006.
28. Quoted in *El Sur,* 29 March 2007.
29. Juan Rulfo, *Pedro Páramo* (New York: Grove, 1994), p. 3.
30. See PDPR-CGEPR, "Un Poco Mas de Historia," September 2005, http://usuarios.lycos.es/cedema/Mexico/EPR/epr(2005septiembre).html, pp. 7–8, viewed 7 October 2005.
31. See Ramon Miro, "Organized Crime and Terrorist Activity in Mexico, 1999–2002," Federal Research Division, U.S. Library of Congress, Washington, DC, February 2003.
32. Interview by author with Maribel Gutiérrez, a reporter (with *El Sur* in Acapulco), Acapulco, Guerrero, 19 May 2004.
33. Interview by author with Hilario Mesino, Founder, Organización de Campesinos de Sierra del Sur (OCSS), Atoyac de Álvarez, Guerrero, 25 May 2003. Also, statements by Julio Mata (President), Tita Radilla (Vice President), and Hilario Mesino (Founder), OCSS, testimony before the UN Human Rights High Commissioner for Human Rights, Chilpancingo, 19 May 2003 (quotes recorded by author who was present at the proceedings).
34. Comandancia General del Ejército Popular Revolucionario, "Manifesto de Aguas Blancas," 28 June 1996, http://burn.ucsd.edu/~cpa/vozfronteriza/oct96/pg08.html, viewed 9 September 1998.
35. Ibid.
36. EPR, "Manifesto de la Sierra Madre Oriental," 7 August 1996, www.pengo.it/PDPR-EPR/manif_smo.htm, viewed 9 September 1998.
37. Ibid.
38. Quote from Claudio Albertani, "Cronología minima de una Guerra Social," January 2004, on the EPR's Web site, www.geocities.com/Bunker/5061/chrono6, viewed 13 March 2004.

39. Quoted in *La Jornada*, 25 August 1996.
40. Interview by author with Juan Sosa Maldonado, Jefe, Organización Pueblos Indígenas Zapotecas, Oaxaca, 12 May 2003.
41. Interview by author with Yésica Sánchez Maya, Directora, Liga Mexicana de Derechos Humanos, Oaxaca, 13 May 2003. Interview by author with Ronaldo González Espinosa, Secretaria Técnico de la Red Oaxaqueña de Derechos Humanos, Oaxaca, 14 May 2003, who indicated that besides Loxicha, significant government repression occurred in other regions such as San Jacinto Tlacotepec, Santiago Amoltepec, Santiago Xochiltepec, Santiago Textitlan, and Santa Maria Zanisa.
42. EPR, "Documento de la Organización, #2," 1996, http://usuarios.lycos.es/cedema/Mexico/EPR/epr(doc2).html, viewed 3 February 2000.
43. The peso crisis was rooted in part in the perception by global investors that instability was rife in Mexico, as exemplified by the presence of guerrillas such as the Ejército Zapatista de Liberación Nacional (EZLN).
44. See Mark Wrighte, "The Real Mexican Terrorists: A Group Profile of the Popular Revolutionary Army," *Studies in Conflict and Terrorism*, vol. 25, #1, May 2002, p. 211.
45. Interview by author with Abel Barrera Hernández, Director, Centro de Derechos Humanos Tlachinollan (CDHT), Tlapa de Comonfort, Guerrero, 21 May 2003.
46. Alejandro Martínez Carvjal, *Ejército Popular Revolucionario* (Mexico, DF: Editorial Sagitario, 1998), p. 254.
47. Laura Castellano, "Tres Décades de Contra Insurgencia en Guerrero," *La Jornada*, 25 January 1998.
48. *La Jornada*, 12 March 1997. This entailed the testimony of Jorge Salas Dircio to various human rights agencies.
49. See Castellano, "Tres Décades de Contra Insurgencia en Guerrero."
50. For a broader discussion of this, see Maribel Gutiérrez, *Violencia en Guerrero* (Mexico, DF: Demos, 1998), pp. 304–306.
51. See Liga Mexicana por la Defensa de los Derechos Humanos, "Informe Sobre la Masacre de El Charco," 7 June 1999, www.derecho.org/limedddh/informes/charco.html, viewed 2 January 2003.
52. Ibid.
53. See David Pavón, "Cronología del EPR y del EPRI," *Revolución*, #25, June 2000, www.geocities.com/Pentagon/Bunker/5061/cron3.htm/200520, viewed 4 April 2004.
54. Ibid. See also Joshua Paulson, "Rural Rebellion in Southern Mexico," *Nacla Report on the Americas*, vol. 33, #5, March 2000, pp. 4–5.
55. *Proceso*, #1200, 31 October 1999.
56. David Pavón and Maria Luisa Vega, "Cronología del Ejército Popular Revolucionario (EPR) y las Fuerzas Armadas Revolucionarias del Pueblo (FARP)," *Revolución*, #27, 2 May 2000, www.geocities.com/Pentagon/Bunker/5061/cron4.html, viewed 3 March 2002.
57. EPR, "Comunicado Sobre Las Elecciones del 2 de Julio," www.pengo.it/PDPR-EPR/comunicados/c_180500.htm, viewed 18 June 2002.
58. EPR, "Comunicado Revolucionario: Posicion Política ante el Nuevo Gobierno de V. Fox," 28 August 2000, www.pengo.it/PDPR-EPR/comunicados/c_280800.htm, viewed 23 August 2003.
59. EPR, "Nuestra Estrategia y Tactica Militar," July 2000, http://usuarios.lycos.es/cedema/Mexico/EPR/epr(estrategia).html, viewed 15 February 2001.
60. Quoted in *Proceso*, #1225, 23 April 2000.
61. *El Sur*, 28 May 2003.
62. Interview by author with Abel Barrera Hernández.

63. EPR, Comite Central del Partido Popular Revolucionario, "Al Pueblo Iraqi," 11 April 2003, www.geocities.com/Pentagon/Bunker/5061/irak.html/200520, viewed 3 May 2004.
64. Interview by author with an official at Centro de Información de Seguridad Nacional (CISEN), anonymity requested, Mexico City, 14 May 2004.
65. Interview by author with Maribel Gutiérrez.
66. Interview by author with Francisco Díaz Gonzalez, Centro de Derechos Humanos Mahatma Gandhi, Coyuca de Benítez, Guerrero, 24 May 2004.
67. EPR, "Comunique," 13 April 2005, www.pdpr-epr.org, viewed 12 November 2005.
68. These comments came up in a dialogue I had with students and faculty when presenting a paper at the Universidad Autónoma de Guerrero, 12 May 2005.
69. PDPR-CGEPR, "Un Poco Mas de Historia," September 2005, http://usuarios.lycos.es/cedema/Mexico/EPR/epr(2005septiembre).html, viewed 12 November 2005.
70. Ibid.
71. See the EPR's *El Insurgente,* vol. 9, #85, December 2005, www.pdpr-epr.org, viewed 6 March 2006.
72. This point was developed in the previous chapter that addressed the political economy of social struggle.
73. Che Guevara, *Guerrilla Warfare* (Lincoln: University of Nebraska Press, 1998), p. 10.
74. See, for example, Guevara, *Guerrilla Warfare,* pp. 34–35.
75. Juan Rulfo, "Paso del Norte," in *The Burning Plain and Other Stories* (Austin: University of Texas Press, 2003), p. 121.
76. The comparison does not include the drug trade.
77. Figures are from the World Bank, and are published in *La Jornada,* 25 January 2006.
78. This is based on an estimate from Mexico's Consejo Nacional de Población, and it appeared in *La Jornada,* 28 December 2005.
79. Figures are from INEGI, as published in *El Sur,* 13 December 2004.
80. The economically active portion of Guerrero's population in 2000 was 1,233,746. INEGI, "Población de 12 años y más por condición de actividad económica, 2004," www.inegi.gob.mx.
81. INEGI, "Porcentaje de población emigrante a Estados Unidos de América por entidad federativa según sexo, 2000," www.inegi.gob.mx.
82. Representante de Organizaciones No-gubernamentales de Derechos Humanos, de Organizaciones Indígenas de Violaciones de los Derechos Humanos, "Informe Sobre la Situacion de los Derechos Humanos en Guerrero," mimeograph, presented in Chilpancingo, Guerrero, 19 May 2003, p. 2.
83. A current tabulation of the number of Guerrero migrants killed en route to or in the United States can be found at the Web site for the human rights group CDHT in Tlapa de Comonfort, Guerrero, www.tlachinollan.org.
84. Miro, "Organized Crime and Terrorist Activity in Mexico," p. 21.
85. Interview by author with Ana Soto, Coordinadora de Derechos Humanos Fray Tapias de Cordova, Tapachula, Chiapas, 3 June 2004; and Abel Barrera Hernández.
86. As quoted in *La Jornada,* 27 January 2006.
87. Remittances grew just 5.4 percent during the first two months of 2007 over 2006, compared to a 27.6 percent growth rate for the same months between 2005 and 2006. See *La Jornada,* 31 March 2007.
88. Carlos Fuentes, *The Old Gringo* (New York: Noonday Press, 1985), p. 5.
89. "Capitol Hill Hearing Testimony," "US-Mexico Border Security—Statement of David Aguilar, Chief, Border Patrol, US Customs and Border Protection," hearing before the U.S. Committee on Senate Judiciary Subcommittee on Immigration, Border Security,

and Citizenship, *Congressional Quarterly,* Federal Document Clearing House Congressional Testimony, 7 June 2005.

90. *La Jornada,* 29 November 2005. See also *New York Times,* 9 January 2005.
91. As quoted in the *Globe and Mail,* 25 March 2006.
92. Testimony of David Aguilar, see note 89 above.
93. Testimony of Joel S. Gallay, Deputy Inspector General, U.S. General Services Administration, before the House Homeland Security Subcommittee on Management, Integration, and Oversight, "Mismanagement of Border Surveillance System," *Congressional Quarterly,* 16 June 2005.
94. As reported by the *Buffalo News,* 18 June 2005.
95. Interview by author with a high-placed representative from the U.S. Department of Homeland Security, anonymity requested, Washington, DC, 30 July 2004.
96. See *Globe and Mail,* 25 March 2006; and *New York Times,* 1 May 2006.
97. Quoted in the *New York Times,* 15 December 2005.
98. Data from the National Network for Immigrant and Refugee Rights, www.nnirr.org/news/press_releases/2005_10_31.htm, viewed 31 October 2005.
99. Quoted in *La Jornada,* 22 June 2005.
100. Interview by author with two high-ranking members of CISEN. These were conducted on 14 May 2004 and 6 May 2005, both in Mexico City. Anonymity requested by officials.
101. *Zona Libre,* Tapachula, Chiapas, Mexico, 3 June 2004. The story is based on an interview with Mauricio Gándara of the Instituto Nacional de Migración.
102. See Josiah Heyman, "Ports of Entry as Nodes in the World System," *Identities,* vol. 11, 2004, pp. 303–327, especially p. 322.
103. *La Jornada,* 9 January 2005. The statistics are from the Comisión Económica para América Latina y el Caribe (Cepal).
104. Rulfo, *Pedro Páramo,* p. 68.
105. See Fuentes, *Old Gringo,* p. 185.
106. See Friedrich Nietzsche, "Good and Evil, Good and Bad," in *The Genealogy of Morals* (New York: Random, 1989), pp. 24–56.

Chapter 7

1. for a broader discussion of the historical aspects of Mexican narcotrafficking, see James Rochlin, *Redefining Mexican Security: Society, State and Region under Nafta* (Boulder, CO: Lynne Rienner, 1997), chapter 4.
2. For an excellent general discussion of this, see John Bailey and Jorge Chabat, eds., *Crimen Transnacional y Seguridad Pública* (Mexico, DF: Plaza y Jánes, 2003).
3. As reported in the *Tampa Tribune,* 1 August 2005.
4. See Ramon Miro, "Organized Crime and Terrorist Activity in Mexico, 1999–2002," U.S. Library of Congress, Federal Research Division, Washington, DC, February 2003, p. 1; and "The Americas: War without End, Drugs in Mexico," (no author), *The Economist,* #8365, 6 March 2004, p. 56.
5. As reported in *La Jornada,* 4 August 2005.
6. See *El Sur,* 5 July 2004.
7. Interview by author with Maribel Gutiérrez Morena, Reportera, *El Sur,* Acapulco, 19 May 2004.
8. See, for example, *El Sur,* 19 April 2004, 20 April 2005, and 23 January 2006.
9. See *La Jornada,* 6 August 2005.

10. *El Sur,* 14 February 2006.
11. Caracol Radio, 12 June 2005, www.caracol.com.co, viewed 12 June 2005.
12. See *La Jornada,* 7 August 2005.
13. *Los Angeles Times,* 24 March 2007.
14. *New York Times,* 27 January 2006.
15. Ibid.
16. *La Jornada,* 18 February 2007.
17. As reported in the *New York Times,* 10 February 2006.
18. *New York Times,* 7 February 2005.
19. *Reforma,* 19 May 2003.
20. Due to the sensitive nature of such observations, and the brutal retaliation to which those persons might be subjected, their identities are kept anonymous here.
21. *El Sur,* 27 January 2006.
22. As reported in *La Jornada,* 12 June 2005.
23. As reported in *El Sur,* 24 May 2003.
24. This point was made by two members of the OCSS while testifying before the UN High Commission of Human Rights, Chilpancingo, Guerrero, 19 May 2003.
25. *El Sur,* 6 June 2005.
26. For a broader discussion of this, see Jordi Pius Llopart, "Guerrero Ecologists Promote Drug Crop Alternatives," *Nacla Report on the Americas,* vol. 36, #2, September 2002, pp. 20–21.
27. Interview by author with, among others, Francisco Díaz González, Centro de Derechos Humanos Mahatma Gandhi, Coyuca de Benítiz, Guerrero, 24 May 2004; Maribel Guitiérrez Morena, Acapulco; and Hilario Mesino Acosta, OCSS, Atoyac de Álvarez, Guerrero, 25 May 2004.
28. For a broader discussion of this, see Rebecca Overmyer-Velasquez, "The Self-determination of Indigenous Peoples and the Limits of United Nations Advocacy in Guerrero, Mexico, 1998–2000," *Identities,* vol. 10, #9, 2003, pp. 10–12.
29. As reported in *La Jornada,* 21 April 2005.
30. "Kidnapping Increases in Latin America," *Janes Intelligence Review,* 1 June 2003, www.janes.com, viewed 2 June 2003.
31. See *La Jornada,* 3 August 2005.
32. A good historical overview of U.S. interests in Mexican oil can be found in Richard Fagen, "Mexican Petroleum and US National Security," *International Security,* vol. 4, #1, 1979, pp. 39–53.
33. As reported in the *Philadelphia Inquirer,* 19 March 2006.
34. Statistics as reported by PEMEX in *La Jornada,* 30 March 2006.
35. As reported in *The Herald,* 19 April 2005.
36. Carmen Gentile, United Press International, "Mexico Faces Production Decline," 15 February 2006.
37. See *La Jornada,* 7 and 8 November 2004.
38. Centro de Estudios Superiores Navales.
39. *El Universal,* 30 January 2006.
40. For more about the Plan Puebla Panama (PPP) from the perspective of the Inter-American Development Bank, see www.iadb.org/aboutus/II/re_ppp.cfm?language=English.
41. Centro de Investigaciones Económicos y Políticas de Acción Comunitaria (CIEPAC), "The La Parota Dam, Resistance in Guerrero," *Chiapas al Día,* #399, 10 March 2004.
42. *La Jornada,* 15 December 2005.

43. *El Sur,* 30 January 2006. The death in January 2006 involved a member of the Consejo de Ejidos y Comunidades Opositores a la Presa La Parota (Cecop).
44. *La Jornada,* 17 December 2005.
45. As reported in *El Sur,* 2 June 2006.
46. Quote from Subcomandante Marcos, as reported in *El Sur,* 16 April 2006.
47. For other comparative data, see Instituto Nacional de Estadísticas Geografia y Informática, www.inegi.gob.mx.
48. Quotes from Marcos as reported in *El Sur,* 17 April 2006.
49. This includes presidents of Belize, Guatemala, Costa Rica, Honduras, El Salvador, and Panama.
50. For a good critical overview of the plan, see Imtiaz Hussain, "Of Mountains, Molehills and Mirages: Canadian and Mexican Central American Pursuits," *Revista Mexicana de Estudios Canadienses,* vol. 1, #7, November 2004, pp. 1–10; and Alejandro Alvarez Béjar, "The Puebla Panama Plan," www.redem.buap.mx/word/alejandro3.doc, viewed 25 November 2004.
51. See Inter-American Development Bank, "Plan Puebla-Panama," www.iadb.org/ppp/pppdescription.asp, viewed 10 February 2006.
52. See Miquel Pickard, "The Plan Puebla Panama Revived," pamphlet, Americas Program, Interhemispheric Resource Center, Silver City, New Mexico, 8 June 2004.
53. Interview by author with Rolando González Espinosa, Secretaria Técnico de la Red Oaxaqueña de Derechos Humanos, Oaxaca, 14 May 2003; Javier Balderas Castillo, Director, Centro de Derechos Humanos Tepeyac, Tehuantepec, Oaxaca, 30 May 2003; and Cirino Placido Valerio and Zosima Avilez Mendoza, Coordinadora Regional de Autoridades Comunitarias (CRAC), San Luis Acatalán, Guerrero, 9 May 2005.
54. See, for example, Centro de Derechos Humanos Tepeyac del Istmo de Tehuantepec, "Intereses y Resistencias: El Plan Puebla Panama y el Corredor Carretero Oaxaca-Istmo-Huatulco," mimeograph, Oaxaca, September 2002, pp. 1–5.
55. See Pickard, "Plan Puebla Panama Revived."
56. See *La Jornada,* 6 February 2006.
57. See Wendy Call, "PPP Focus Moves South as Mexican Backing Loses Momentum," PPP Spotlight #1, Interhemispheric Resource Center, Silver City, New Mexico, 20 February 2003.
58. See Rochlin, *Redefining Mexican Security,* pp. 141–146.
59. Interview by author with Juan Sosa Maldonado, Jefe, Organización Pueblos Indígenas Zapotecas, Oaxaca, 12 May 2003 and 28 May 2004; Yésica Sánchez Maya, Directora, Liga Mexicana de Derechos Humanos, Oaxaca, 12 May 2003; Rolando González Espinosa, Oaxaca, 14 May 2003 and 31 May 2004; Dr. Victor Raul Martínez, Sociologist, Universidad Autónoma Benito Juárez de Oaxaca, Oaxaca, 16 May 2003; Javier Balderas Castillo; and Minerva Nora Martínez Lázaro, Directora, Centro Regional de Derechos Humanos Bartoleme Carrascpo, Oaxaca, 27 May 2004.
60. See *La Jornada,* 12 December 2006.
61. Interview by author with Hilario Mesino Acosta, Guerrero, 23 May 2003. Other members were present during the interview.
62. Ibid.
63. Interview by author with Hilario Mesino Acosta, Guerrero, 25 May 2004.
64. This was the case through early 2006.
65. These include the Instituto Guerrerense de Derechos Humanos in Chilpancingo, the Comisión de Derechos Humanos Mahatma Gandhi in Coyuca de Álvarez, Centro Regional de Defensa de Derechos Humanos Jose Maria Morelos y Pavón in Chilapa de

Álvarez, the Comisión de Derechos Humanos Sembrador de la Esperanza in Acapulco, and the CDHT in Tlapa de Comonfort.

66. See "Informe Sobre la Situación de los Derechos Humanos en Guerrero," Representante de Organizaciones No-Gubernamentales de Derechos Humanos de Organizaciones Indígenas de Violaciones de los Derechos Humanos, 19 May 2003.
67. Interview by author with Francisco Díaz González.
68. The illiteracy rate is as reported in *El Sur,* 14 June 2004.
69. The figure is from INEGI and is reported in *El Sur,* 31 May 2006.
70. Interview by author with Abel Barrera Hernández, Director, CDHT, Tlapa de Comonfort, Guerrero, 21 May 2003; and Leslie Davies, International Coordinator, CDHT, Tlapa de Comonfort, Guerero, 18 May 2004.
71. See, for example, the Web site of CDHT for the group's assessment of key security themes facing Guerrero, www.tlachinollan.org.
72. Michel Foucault, *Society Must Be Defended* (New York: Picador, 1997), p. 29.
73. EZLN, Clandestine Revolutionary Indigenous Committee, "Sexta Declaración de la Selva Lacandona," June 2005, www.ezln.org/documentos/2005/sexta.es.htm, viewed 6 July 2005.
74. Michel Foucault, "The Birth of Biopolitics," in *Ethics: Subjectivity and Truth,* ed. Paul Rabinow (New York: New Press, 1997), p. 73.
75. Interview by author with three anonymous and masked members of the Junta de Buen Gobierno de Oventic, Caracol de Oventic, Chiapas, Mexico, 16 May 2005.
76. Ibid. The three Junta members took my written list of questions prior to the interview, and left the room for about one-half hour to discuss them with other members of the Caracol. When they returned, they indicated they could not address those two questions.
77. Carlos Fuentes, *The Death of Artemio Cruz* (New York: Farrar, Straus and Giroux, 1991), p. 187.
78. Interview by author with Cirino Placido Valerio (of the Policías Comunitarias in San Luis Acatlán) and Zosimo Avilez Mendoza (of the original Policías Comunitarias in Santo el Rincón), both of the CRAC, San Luis Acatlán, Guerrero, 9 May 2005.
79. Ibid.
80. There exists no other data to confirm or deny these statistics.
81. National Commission for the Development of Indigenous Peoples.
82. Ibid.
83. Comite Ejectivo de la Policía Comunitaria, San Luis Acatlán, "Sobre el Emplazamiento que el Gobierno Guerrerense ha hecho Nuevamente al Sistema Comunitario de Seguridad e Impartición de Justicia para su Legalización," mimeograph, San Luis Acatlán, Guerrero, 2 July 2004.
84. As reported in *El Sur* (Acapulco), 17 October 2005.
85. *El Sur* (Acapulco), 4 June 2005.
86. CRAC, "10 Aniversario de la Policía Comunitaria—Mesas de Trabajo y Declaracion Final," October 2005, available at Web site of CDHT, www.tlachinollan.org, viewed 4 November 2005.
87. Interview with CRAC leaders, see note 78 above.
88. Interview by author with Abel Barrera Hernández, Guerrero, 31 May 2003. This human rights agency is as strong and adept as any in Chiapas, and deals with the areas of Guerrero that are affected by Policías Comunitarias. This movement is not addressed significantly in the academic literature, although scant attention exists. An example is a very brief reference made to them by Carlos García, "Inventario de las Organizaciones Campesinas," in *Crónicas del sur: Utopías campesinas en Guerrero,* ed. Armando Bartra (Mexico, DF: Ediciones Era, 2000), p. 115.

89. Fuentes, *The Death of Artemio Cruz,* p. 103.
90. See *El Norte* (Monterrey, Mexico), 12 November 2006.
91. See, for example, Karl Marx, *The 18th Brumaire of Louis Bonaparte* (New York: International, 2004), pp. 35–37. See also Barrington Moore, Jr., *The Social Origins of Dictatorship and Democracy: Lord and Peasant in the Making of the Modern World* (Boston: Beacon Press, 1966).
92. For information on the "Other Campaign," see their Web site: www.ezln.org.
93. Interview by author with Dr. Raul Benítez, Investigador, Centro de Investigaciones Sobre America del Norte, Universidad Autónoma Nacional de México (UNAM), Mexico City, 7 May 2003.
94. Interview by author with Hilario Mesino Acosta, 23 May 2003; Abel Barrera Hernández, 31 May 2003; Yésica Sánchez Maya; Rolando González Espinosa; Juan Sosa Maldonado; and Francisco Díaz González.
95. Interview with Jaime Salazar, Departmento de Historia, Universidad Autonoma Nacional, Chilpancingo, Guerrero, 20 May 2004.
96. Interview with Abel Barrera Hernández, 31 May 2003.
97. For a discussion of the feud, see *La Jornada,* 16 April 2006.
98. Interview by author with Rolando González Espinosa; Juan Sosa Maldonado; Yesica Sánchez Maya; Victor Raul Martínez; Javier Balderas Castillo; and Minerva Nora Martínez Lázaro.

Chapter 8

1. Miguel Cervantes, *Don Quixote* (New York: Oxford University Press, 1999), p. 1115.
2. See Sun Tzu, *The Art of War* (Boston: Shambhala, 1991), p. 49.
3. Ralph Sawyer, ed., *The Seven Military Classics of Ancient China* (Boulder, CO: Westview, 1993).
4. Carlos Fuentes, *The Death of Artemio Cruz* (New York: Farrar, Straus and Giroux, 1991), p. 56.
5. For example, an AUC leader, Salvatore Mancuso, admitted paying the Colombian police and military forces some $400,000 (US) for their cooperation in key oil-producing areas such as Catatumbo, in addition to an unfolding slew of other evidence linking the AUC to prominent members of the Colombian government since at least the 1990s. See, for example, *El Tiempo,* 16 January 2007 and 15 January 2007.
6. See *Los Angeles Times*, 22 January 2007.
7. *La Jornada,* 6 April 2007.
8. Hannah Arendt, *On Violence* (New York: Harvest, 1970), p. 44.
9. Interview by author with Daniela Oyague, Trade Commissioner, Canadian Embassy, Caracas, 8 August 2006.
10. See James Rochlin, *Vanguard Revolutionaries in Latin America: Peru, Colombia and Mexico* (Boulder, CO: Lynne Rienner, 2003).
11. Interview by author with Juan Ramon Rios, Director, Secretario General, USO, Bogotá, 14 July 2006.
12. Interview by author with Tarsicio Mora Fodoy, Fiscal de Central Unitaria de Trabajadores de Colombia, Bogotá, 12 July 2006.
13. Michel Foucault, *The Order of Things* (New York: Vintage, 1994), p. 50.
14. For example, Athens' Pericles admits that Athens' colonial empire is based on tyranny and cites this as a source of strategic worry (Thucydides, *History of the Peloponnesian*

War, p. 126). In another of many examples, Thucydides notes that "so general was the indignation felt against Athens, whether by those who wishe[d] to escape from her empire, or those who were apprehensive of being absorbed by it" (pp. 93–96). See Thucydides, *A History of the Peloponnesian War,* ed. Robert Strassler (New York: Free Press, 1996).

Bibliography

Abdel-Fatua and J. Kayode Fayemi, eds., *Mercenaries: An African Security Dilemma* (London: Pluto, 2000).

Adams, Thomas, "The Real Military Revolution," *Parameters,* vol. 30, #3, Autumn 2000, pp. 54–65.

Aguayo Quezada, Sergio, *La Charola: Una historia de los servicios de inteligencia en Mexico* (Mexico, DF: Grijalbo, 2001).

Almario Salazar, Gustavo, *Historia de los Trabajadores Petroleros* (Bogotá: Ediciones Cedetrabajo, 1984).

Álvarez, Victor, "Notas Sobre la Constitución del Estado en Colombia," in *El Estado y la Fuerza,* ed. Freddy Salazar (Bucaramanga: Universidad Industrial de Santander, 1999), pp. 121–176.

Amnesty International, "Colombia: Just What Do We Have to Do to Stay Alive?" 1997, www.amnesty.org/library/print/ENGAMR230481997, viewed 3 October 2003.

Apuleyo Mendoza, Plinio, "Las FARC: Un Enclave Terrorista en Colombia," in *Terrorismo y Seguridad,* ed. Reinaldo Botero (Bogotá: Planeta, 2003), pp. 195–210.

Arnold, Guy, *Mercenaries: The Scourge of the Third World* (New York: St. Martins, 1999).

Arquilla, John, "Strategic Implications of Information Dominance," *Strategic Review,* vol. 22, #3, Summer 1994, pp. 24–30.

Arquilla, John, and David Ronfeldt, "Cyberwar Is Coming!" *Comparative Strategy,* vol. 12, #2, 1993, pp. 141–165.

Autodefensas Unidas de Colombia (AUC), "27 Años de lucha por la paz," February 2005, www.colombialibre.org/detalle_col.php?banner=editorial&id=10507, viewed 20 March 2005.

———, "Agenda de Negociación AUC," 25 May 2004, www.salvatoremancuso.com/detalled.php, viewed 22 May 2006.

———, Carlos Castaño, "Carta Abierta a Andrés Pastrana Arango, Presidente de la Republica," 17 February 2001, www.colombialibre.org/cartas/Pastrana_17.htm, viewed 2 April 2003.

———, "Comunicado a la Opinion Nacional e Internacional," 5 September 2004, www.colombialibre.org, viewed 12 November 2004.

———, "Declaración por la Paz en Colombia," 29 November 2002, www.salvatoremancuso.com/detalle.php, viewed 10 May 2006.

———, "Discurso del Jefe del Estado Mayor de las AUC, Comandante Salvatore Mancuso, en el acto de Instalacion Oficial del Proceso de Negociación entre el

Gobierno Nacional y las Autodefensas Unidas de Colombia," July 2004, www.colombialibre.org/ver_imp.php/Varid=7718, viewed 18 October 2005.

———, "El accompañamiento: Indespendable para la post-demovilización," 27 April 2006, www.colombialibre.org/detalled_col.php?banner=Editorial&nreg=0, viewed 10 May 2006.

——, "Entrevista a Los Comandantes de Las ACCU: Santander Lozada y Cesar Marin," 9 and 10 September 1998, www.colombialibre.org/santa.htm, viewed 2 April 2003.

———, Carlos Castaño, "Entrevista del Comandante Castaño a RCN Radio," 18 February 2001, www.colombialibre.org/reportajes/rcnradio.htm, viewed 13 September 2002.

———, "Intervención del Alto Comisionado para La Paz, Luis Carlos Restrepo," 12 April 2005, www.salvatoremancuso.com/detalle.php, viewed 21 May 2006.

———, Estado Mayor Negociador, Salvatore Mancuso, "Nuestras Propuesta de Incorporación a las Fuerzas Armadas," 26 October 2004, www.colombialibre.org/ver_imp.php?Varid=8398, viewed 10 November 2004.

———, "Quién debe ganar la Guerra?" 24 April 2003, www.salvatoremancuso.com/dettalle.php, viewed 22 May 2006.

———, "Reportaje del Comandante Castaño el periodico Notitarde," 23 December 2000, www.colombialibre.org/reportajes/notitarde.htm, viewed 2 January 2002.

———, "Triumfo de Uribe," 1 June 2006, www.colombialibre.org/detalled_col.php, viewed 2 June 2006.

Avilés, William, *Global Capitalism, Democracy, and Civil-Military Relations in Colombia* (Albany: SUNY, 2006).

Bacon, David, "The Most Dangerous Place to Join a Union," *Progressive,* vol. 67, #11, 2003, pp. 29–31.

Bailey, John, and Jorge Chabat, eds., *Crimen Transnacional y Seguridad Pública* (Mexico, DF: Plaza y Jánes, 2003).

Barr, Bob, "The War without End—Colombia," *American Spectators,* vol. 37, #3, 2003, pp. 26–36.

Bartra, Armando, *Cronícas del sur: Utopías campesinas en Guerrero* (Mexico, DF: Ediciones Era, 2000).

———, *Guerrero Bronco: Campesinos, ciudadanos y guerrilleros en la Costa Grande* (New Haven, CT: Yale University Press, 1996).

Baudrillard, Jean, *The Illusion of the End* (Palo Alto, CA: Stanford University Press, 1992).

Bentham, Jeremy, *The Panopticon Writings,* ed. M. Bozovic (London: Verso, 1995).

Berkowitz, Bruce, "Better Ways to Fix US Intelligence," *Orbis,* vol. 45, #4, Fall 2001, pp. 609–620.

Berquist, Charles, *Labor and Latin America: Comparative Essays on Chile, Argentina, Venezuela and Colombia* (Palo Alto, CA: Stanford University Press, 1986).

Berry, Albert, "The Income Distribution Threat in Latin America," *Latin American Research Review,* vol. 32, #2, 1997, pp. 3–40.

Betts, Richard, "Fixing Intelligence," *Foreign Affairs,* vol. 18, #1, January 2002, pp. 43–60.

Bjorgo, Tore, ed., *Terror from the Extreme Right* (London: Frank Cass, 1995).
Black, Jeremy, "War and Strategy in the 21st Century," *Orbis,* vol. 46, #1, Winter 2002, pp. 137–144.
Bland, Douglas, "The RMA: Managing an Idea," in *Managing the Revolution in Military Affairs,* ed. Ron Matthews (New York: Palgrave, 2001), pp. 19–35.
Bonilla Machorro, Carlos, *Ejercicio de Guerrillero* (Mexico, DF: Gaceta, 1981).
Botero, Reinaldo, ed., *Terrorismo y Seguridad* (Bogotá: Planeta, 2003).
Brooks, Doug, "Messiahs or Mercenaries?" *International Peacekeeping,* vol. 7, #4, Winter 2000, pp. 129–144.
Browitt, Jeff, "Capital Punishment: The Fragmentation of Colombia and the Crisis of the Nation-State," *Third World Quarterly,* vol. 22, #6, 2001, pp. 1063–1076.
Cabañas, Lucio, *Lucio Cabañas, el guerrillero sin esperanza,* ed. Luis Suárez (Mexico, DF: Roca, 1978).
Caferro, William, *Mercenary Companies and the Decline of Siena* (Baltimore: Johns Hopkins University Press, 1998).
Caicedo, Edgar, *Historia de las luchas sindicales en Colombia* (Bogotá: Ediciones CEIS, 1971).
Cameron, Maxwell, "Strengthening Checks and Balances—Democracy Defence and Promotion in the Americas," *Canadian Foreign Policy,* vol. 10, #3, 2003, pp. 101–116.
Campen, Alan, Douglas Dearth, and Thomas Gooden, eds., *Cyberwar: Security, Strategy, and Conflict in the Information Age* (Fairfax, VA: Armed Forces Communications and Electronics Association, 1996).
Campos Gómez, Eleazar (Comandante), *Lucio Cabañas y el Partido de los Pobres* (Mexico, DF: Editorial Nuestra America, 1987).
Center for International Policy, "Colombia Program—Military and Police Assistance Programs," www.ciponline.org/colombia/aidtable.htm, viewed 3 May 2006.
———, "US Support for Plan Colombia," 2 September 2003, www.ciponline.org/facts/coaid.htm, viewed 13 October 2003.
Centro de Derechos Humanos Tepeyac del Istmo de Tehuantepec, "Intereses y Resistencias: El Plan Puebla Panama y el Corredor Carretero Oaxaca-Istmo-Huatulco," mimeograph, September 2002.
Cervantes, Miguel, *Don Quixote* (New York: Oxford University Press, 1999).
Chapman, Bert, "Revolution in Military Affairs," *Bulletin of Bibliography,* vol. 59, #1, 2002, pp. 1–25.
Chatterjee, Pratap, "Guarding the Multinationals," *Multinational Monitor,* vol. 19, #3, March 1998, pp. 1–5.
Chernick, Marc, "La Negociación de Un Paz Entre Múltiples Formas," in *Los Laberintos de La Guerra,* ed. Francisco Leal Buitrago (Bogotá: Tercer Mundo, 1999), pp. 3–58.
Clausewitz, Carl von, *On War* (Princeton, NJ: Princeton University Press, 1989).
Coatsworth, John, "Colombia: Roots of Violence in Colombia," *Revista: Harvard Review of the Americas,* vol. 2, #3, Spring 2003, pp. 1–4.
Cohen, Nick, "The Curse of Black Gold," *New Statesman,* vol. 132, 2003, pp. 25–28.
Comisión Internacional de Juristas, "Colombia: Políticos del Gobierno Socavan el

Estado de Derecho y Consolidan la Impunidad," mimeograph, Bogotá, 21 September 2005.

Consultoría para los Derechos Humanos y el Desplazamiento Forzado (Codhes), "Comportamiento de Desplazamiento, 1985–2005," www.codhes.org.co/cifra/GraficoTendencias1985_2005.jpg, viewed 10 May 2006.

———, *Un Pais Que Huye: Desplazamiento y violencia* (Bogotá: Codhes, 2003).

Corporacion Andina de Fomento, *Venezuela y Colombia en el Nuevo Milenio* (Bogotá: Fundacion Pensamiento y Accion, 1998).

Crandall, Russell, *Driven by Drugs: United States Policy toward Colombia* (Boulder, CO: Lynne Rienner, 2002).

Crenshaw, Martha, ed., *Terrorism in Context* (University Park: Pennsylvania State University Press, 1995).

Dartness, Michael, "Insurgency Online," *Small Wars and Insurgencies,* vol. 10, #3, Winter 1999, pp. 128–143.

Davenport-Hines, Richard, *The Pursuit of Oblivion: A Global History of Narcotics, 1500–2000* (London: Weidenfeld and Nicolson, 2001).

Deas, Malcolm, and Fernando Gaitán Daza, *Dos Ensayos Especulativos Sobre la Violencia en Colombia* (Bogotá: Tercer Mundo, 1985).

Defensoria del Pueblo, Colombia, Secretariado Nacional de Pastoral Social Caritas Colombiana, *Nordeste* (Bogotá: Defensoria del Pueblo, 2000).

de la Peña, Moisés, *Guerrero económico* (Chilpancingo: Gobierno del estado de Guerrero, 1949).

Delgado, Álvaro, *Politica y movimiento obrero, 1970–1983* (Bogotá: Ediciones CEIS, 1984).

Demchak, Chris, "Technology's Knowledge Burden, the RMA and the IDF: Organizing the Hypertext Organization for Future 'Wars of Disruption'?" *Journal of Strategic Studies,* vol. 24, #2, 2001, pp. 77–147.

Der Derian, "Interview with Paul Virilio—Speed Pollution," *Wired,* May 1996, www.wired.com/archive/4.05/virilio_pr.html.

Derrida, Jacques, *Specters of Marx: The State of the Debt, the Work of Mourning, and the New International* (New York: Routledge, 1994).

Dillon, Michael, and Julian Reid, "Global Liberal Governance: Biopolitics, Security and War," *Millennium,* vol. 30, #1, 2004, pp. 41–46.

DISAM Journal of International Security Assistance Management, "US Assistance to Colombia and the Andean Region," no author, vol. 24, #3, 2002, pp. 35–42.

Dix, Robert, "The Varieties of Revolution," *Comparative Politics,* vol. 15, #8, April 1983, pp. 282–294.

Dokubo, Charles, "An Army for Rent, Private Military Corporations," *Civil Wars,* vol. 3, #2, Summer 2000, pp. 51–64.

dos Santos, Theodore, "The Structure of Dependence," *American Economic Review,* vol. 60, #2, 1970, pp. 231–237.

Dudley, Steven, *Walking Ghosts: Murder and Guerrilla Politics in Colombia* (New York: Routledge, 2004).

Ejército de Liberación Nacional (ELN), "2004—Año de Combate Social," *Revista,* #25, 23 December 2004, www.eln-voces.com/Insurreccion/Nacional.html, viewed 10 January 2005.

———, "Comunicado: Contra la Guerra de Uribe Cumplimos con la Paz," 22 December 2003, www.eln-voces.com/Insurreccion/Anteriores/CM/2003?CM161.htm, viewed 16 February 2004.
———, "Comunicado del ELN—Visita no Grata," 28 August 2000, caobo23@geneva-link.ch, e-mail, viewed 10 September 2000.
———, "Conclusiones del sexto pleno de la direccion nacional del ELN," October 2004, www.eln.voces.com/Correa_del_Magdalena/revista%Unidad%20VI.htm, viewed 10 January 2005.
———, "Con Impunidad, No Hay Paz," 26 May 2006, www.eln-voces.com/Insurreccion/Editorial/html, viewed 29 May 2006.
———, "Correo del Magdalena," no. 162, 15 January 2004, www.eln-voces.com/insurreccion/Anteriores/CM/2004/CM161.htm, viewed 5 January 2005.
———, "Coyuntura Nacional: Arauca: Prioridad del Plan Colombia," 22 December 2003, www.eln-voces.com/Insurreccion/Anteriores/CM/2003/CM161.htm, viewed 22 June 2004.
———, "El Sindicalismo Colombiano: En la Mira de la Democracia Paramilitar," 6 July 2004, www.eln-voces.com/Insurreccion/Nacional3.html, viewed 21 September 2004.
———, "Entrevista al Comandante Antonio Garcia," 21 August 2004, www.eln-voces.com, viewed 13 October 2004.
———, "Entrevista con Nicolás Rodríguez y Antonio García—Carocol de Colombia," 1 January 2000, caobo23@geneva-link.ch, e-mail, viewed 22 March 2001.
———, "Pastrana le Incumple a la Paz," 20 July 2000, caobo23@geneva-link.ch, e-mail, viewed 4 February 2001.
———, "Políticas de Guerra," July 2004, www.eln-voces.com, viewed 13 September 2004.
———, "Por un Gran Acuerdo Nacional, Correo del Magdalena, Órgano Informativo del ELN, II," *Época*, #115, 18 October 1999, caobo23@geneva-link.ch, e-mail, viewed 2 January 2002.
———, "Posición del ELN Sobre la Privatización de ISA-ISAGEN," pamphlet, March 2000.
———, "Pronunciamiento," 18 December 2004, www.eln-voces.org/Correa_del_Magdalena/Pronunciamiento-Trinidad-18-12-04.htm, viewed 5 January 2005.
———, "Quienes Somos?" www.eln.voces.com, viewed 10 January 2005.
———, "A Rechazar el Plan Colombia," 9 December 2000, caobo23@geneva-link.ch, e-mail, viewed 30 December 2000.
———, "Venemos Contra la Paz," 20 May 2000, caobo23@geneva-link.ch, e-mail, viewed 25 May 2000; Ejército Peruano, *Guerra no convencional—Contrasubversion* (Lima: Ejército Peruano, 9 August 1989).
Ejército Popular Revolucionario (EPR), "Al Pueblo Iraqi," 11 April 2003, www.geocities.com/Pentagon/Bunker/5061/irak.html/200520, viewed 3 May 2004.
———, Comandancia General del Ejército Popular Revolucionario, "Manifesto de Aguas Blancas," 28 June 1996, http://burn.ucsd.edu/~cpa/voz/fonteriza/oct96/pg08.html, viewed 9 September 1998.
———, "Comunicado, sobre Gutiérrez Rebollo, confirma que el poder del

narcotrafico esta en el propio estado," 26 February 1997, www.pengo.it/pdpr-epr/comunicados/c_280800.htm, viewed 22 December 2000.
———, "Comunicado Revolucionario: Posicion Política ante el Nuevo Gobierno de V. Fox," 28 August 2000, www.pengo.it/PDPR-EPR/comunicados/c_280800.htm, viewed 23 August 2003.
———, "Comunicado Sobre Las Elecciones del 2 de Julio," July 2000, www.pengo.it/PDPR-EPR/comunicados/c_180500.htm, viewed 18 June 2002.
———, "Comunique," 13 April 2005, www.pdpr-epr.org, viewed 12 November 2005.
———, "Documento de la Organización, #2," 1996, http://usarios.lycos.es/cedema/Mexico/EPR/epr(doc2).html, viewed 3 February 2000.
———, *El Insurgente,* vol. 9, #85, December 2005, www.pdpr-epr.org, viewed 6 March 2006.
———, "Manifesto de la Sierra Madre Oriental," 7 August 1996, www.pengo.it/PDR_EPR/manif_sml.htm, viewed 9 September 1998.
———, "Nuestra Estrategia y Tactica Militar," July 2000, http://usarios.lycos.es/cedema/Mexico/EPR/epr(estrategia).html, viewed 15 February 2001.
———, "Proyecto del pais," 2 December 1996, www.pengo.it/pdpr-epr/proyecto.htm, viewed 31 July 2000.
———, Comité Central del Partido Democrático Popular Revolucionario y Comandancia General del EPR, "Un Poca Mas de Historia," September 2005, http://usarios.lycos.es/cedema/Mexico/EPR/epr(2005setiembre).html, viewed 7 October 2005.
Ejército Zapatista de Liberación Nacional (EZLN), Clandestine Revolutionary Indigenous Committee, Subcomandante Marcos, "Chiapas: The Southeast in Two Winds—a Storm and a Prophecy," August 1992, www.ezln.org, viewed 13 March 1998.
———, "Declaración de la Selva Lacandona," December 1993, www.ezln.org/primera-lacandona.htm, viewed 21 November 1997.
———, "Primera declaración de la realidad," January 1996, www.ezln.org/primera-realidad.htm, viewed 21 November 1997.
———, "Segunda declaración de la Selva Lacandona," 10 June 1994, www.ezln.org/archive/segunda-lacandona.html, viewed 3 September 1997.
———, "Sexta Declaración de la Selva Lacandona," June 2005, www.ezln.org/documentos/2005/sexta.es.htm, viewed 6 July 2005.
Ek, Richard, "A Revolution in Military Geopolitics?" *Political Geography,* vol. 19, #7, 2000, pp. 841–874.
Escobar, Arturo, "Displacement, Development, and Modernity in the Colombian Pacific," *International Social Science Journal,* vol. 55, #175, 2003, pp. 157–168.
Escude, Carlos, *Mercenarios del fin del Milenio* (Buenos Aires: Editorial Belgrano, 1999).
Fagen, Richard, "Mexican Petroleum and US National Security," *International Security,* vol. 4, #1, 1979, pp. 39–53.
Fanon, Franz, *The Wretched of the Earth* (New York: Grove Press, 1963).
Ferris, John, and Michael Handel, "Clausewitz, Intelligence, Uncertainty and the Art of Command in Military Operations," *Intelligence and National Security,* vol. 10, #1, January 1995, pp. 1–58.

Ferris, Stephen, and David Keithly, "Outsourcing the Sinews of War: Contractor Logistics," *Military Review,* vol. 81, #5, September-October 2001, www.uscac.army.mil/CAC/milreview/English/septoct01/indexso01.htm, viewed 3 October 2003.

Foucault, Michel, *The Archaeology of Knowledge and the Discourse on Language* (New York: Tavistock, 1972).

———, "The Birth of Biopolitics," in *Ethics: Subjectivity and Truth,* ed. Paul Rabinow (New York: Free Press, 1994), pp. 73–80.

———, *Discipline and Punish: The Birth of the Prison* (New York: Vintage, 1977).

———, *The Order of Things: An Archaeology of the Human Sciences* (New York: Vintage, 1970).

———, *Power/Knowledge: Selected Interviews and Other Writings, 1972–1977,* ed. Colin Gordon (New York: Pantheon, 1980).

———, *Remarks on Marx* (New York Semiotext(e), 1991).

———, *Society Must Be Defended—Lectures at the Collège France, 1975–1976* (New York: Picador, 1997).

———, "Space, Knowledge and Power," in *Power: Essential Works of Foucault, 1954–1984,* ed. James Faubion (New York: Free Press, 2000), pp. 349–364.

———, "The Subject and Power," in Faubion, *Power,* pp. 326–348.

———, "Truth and Juridical Forms," in Faubion, *Power,* vol. 3, pp. 1–89.

———, "Two Lectures," in Gordon, *Power/Knowledge,* pp. 78–108.

———, "The Will to Knowledge," in *Ethics: Subjectivity and Truth; Essential Works of Foucault,* vol. 1, ed. Paul Rabinow (New York: New Press, 1997), pp. 11–16.

Francis, David, "Mercenary Intervention in Sierra Leone: Providing National Security or International Exploitation?" *Third World Quarterly,* vol. 20, #2, 1999, pp. 319–339.

Freire, Paulo, *Pedagogy of the Oppressed* (New York: Continuum, 1982).

Fuentes, Carlos, *The Burning Mirror: Reflections on Spain and the New World* (New York: Mariner Books, 1999).

———, *The Death of Artemio Cruz* (New York: Farrar, Straus and Giroux, 1991).

———, *The Old Gringo* (New York: Noonday, 1985).

Fuerzas Armadas Revolucionarias de Colombia (FARC), "30 años de las FARC-EP, Texto Completo del Discurso Pronunciado por el Comandante en Jefe de las FARC-EP Manuel Marulanda Vélez," 27 May 1994, www.dix.swix.ch/farcep/Nuestra_historia/30_anos_manuel.htm, viewed 3 December 2004.

———, "36 Años por la Paz y la Soberania Nacional," 20 May 2000, www.farcep.org/Nuestra_historia/36_aniversario_de_las_Farc-EP.html, viewed 29 April 2003.

———, "40 Años de Lucha por la Paz," May 2004, www.farcep.org/aniversario/especial40aniv/text10.html, viewed 10 January 2005.

——— "A la Política Agraria de Mentiras de la Oligarquia, oponemos una efectiva política agraria revolucionaria," March 2000, www.six.swix.ch/farcep/audiencias/reforma/html, viewed 24 April 2003.

———, "Carta pública al pueblo y a los sectores progresistas de Norte-America," April 1998, www.burn.uscd.edu~farc-ep/comunicados, viewed 16 November 1999.

———, "Clinton Go Home," 23 August 2000, www.farcep.org/Comunicados/2000/clinton.html, viewed 23 January 2001.

———, "Comunicado de las FARC-EP al Pueblo de Venezuela," 18 August 2004, www.farc-ep.ch, viewed 17 September 2004.

———, "Contra la Intervencion, Unidad Patriótica y Latinamericana," www.burn.ucsd.edu~farc-ep/comunicados, viewed 27 May 2004.

———, "Declaración Política—la patria esta amenazada," March 2000, www.six.swix.ch/farcep/Pleno/declaracion.html, viewed 22 March 2005.

———, "Empresas criminals avalades por Seguridad Democrática," 11 November 2004, www.farcep.org/novedades/coyuntura/paramilitarismo/noviembre_11_2004.php, viewed 22 March 2005.

———, "Historia de las Conferencias," May 2004, www.farcep.org/aniversario/especial40/text04.html, viewed 5 January 2005.

———, "Las FARC-EP: 30 años de lucha por la paz, democracia y soberanía," May 1994, www.analitica.com/biblioteca/farc/30.asp, viewed 2 May 2005.

———, "Ley 02, Sobre la Tributación," March 2000, www.farc-ep.org/, viewed 12 February 2002.

———, "Los pueblos son las Victimas del Terrorismo de Estado de los Gobierno Capitalistas," 18 July 2003, www.farch-ep.ch.

———, "Manifesto de Movimiento Bolivariano por la Nueva Colombia," 25 March 2000, www.farc-ep.org, viewed 2 March 2002.

———, Manuel Marulanda Vélez, "Texto Completo del Discurso Pronunciado por el Comandante en Jefe de las FARC-EP Manuel Marulanda Vélez," 27 May 1994, www.burn.ucsd.edu~farc-ep/comunicados, viewed 22 September 2002.

———, "Nuestra Historia," 2004, www.farcep.org/aniversario/especial40aniv/textcrono.html, viewed 2 May 2005.

———, "Octava conferencia nacional de las FARC-EP: Programa agrario de los guerrilleros de las FARC-EP," 2 April 1993, www.burn.uscd.edu~farc-ep/comunicados, viewed 19 June 1997.

———, "Paramilitarismo y multinacionales," www.farcep.org/novedades/coyuntura/editorial/mayo_21_2004.php, viewed 10 December 2004.

———, "Plan Colombia: Brazo militar del ALCA, sin resultados," 16 August 2003, www.farcep.org/novedades/coyuntura/plan_colombia/agosto162003.php, viewed 10 January 2005.

———, "Plan Colombia y amanaza para todas," 13 August 2001, www.six.swix.ch/farcep/Articulos/plancolombia.html, viewed 29 December 2004.

———, "Saludo de Ano Nuevo a Los Gobiernos y Pueblos del Mundo," January 2004, www.farc-ep.ch, viewed 6 October 2004.

———, "Saludo de las FARC," January 2006, www.farc-ep.ch/novedades/comunicados/sec2006/enero_01b.phb, viewed 3 March 2006.

———, "Socialismo, Experiencias en el Cambio e Integración," 18 May 2006, www.farcep.org/?node=2.2059.1, viewed 29 May 2006.

Galdi, Theodor, "Revolution in Military Affairs? Competing Concepts, Organizational Responses, Outstanding Issues," paper presented to the Federation of American Scientists, Washington, DC, 11 December 1995.

García Márquez, Gabriel, *Living to Tell the Tale* (New York: Alfred A. Knopf, 2003).

———, *One Hundred Years of Solitude* (New York: Avon, 1970).
Gat, Azar, *The Origins of Military Thought from Englihtenment to Clausewitz* (Oxford: Oxford University Press, 1989).
Gedicks, Al, "Resource Wars against Native Peoples in Colombia," *Capitalism, Nature, Socialism,* vol. 14, #2, June 2003, pp. 98–117.
George, Alexander, ed., *Western State Terrorism* (New York: Routledge, 1991).
Gibson, Robert, "Space Power, the Revolution in Military Affairs," Strategic Research Project paper, U.S. Army War College, Carlisle, PA, 10 April 2001.
Gill, Stephen, ed., *Gramsci, Historical Materialism and International Relations* (Cambridge: Cambridge University Press, 1993).
Gilly, Adolfo, *La revolución interrumpida* (Mexico, DF: Ediciones Era, 1994.
Gleick, James, *Chaos: Making a New Science* (New York: Penguin, 1988).
Gobierno de Colombia, Cámara y Senado de Colombia, "Texto Completo Definitivo Conciliado entre Cámera y Senado del Proyecto de Ley de Justicia y Paz," 21 June 2005, www.salvatoremancuso.com/detalled.php, viewed 2 September 2005.
———, DANE, "Población Económicamente Activa y Número de Sindicalizados," Censo ENS, 2002.
———, Defensoria del Pueblo, Secretariado Nacional de Pastoral Social Caritas Colombiana, *Situacion de los DDHH y del Derecho Humanitario en la Sierra Nevada de Santa Marta* (Bogotá: Defensoria del Pueblo, 2003).
———, Departamento de Planeación, *El Gasto Militar: Desarrollo Teórico y Comparativo Internacional, Informe 2, Justicia y Seguridad* (Bogotá: Departamento Nacional de Planeación, 2000).
———, Ministerio de Defensa, "Descripción del apoyo de Estados Unidos al Plan Colombia," mimeograph, 2001.
———, Ministerio de Defensa, "Los grupos ilegales de Autodefensa en Colombia," December 2000, www.derechoshumanos.gov.co/observatorio/04_publicaciones/, viewed 28 April 2004.
González, Fernán, *Pasado y Presente del Sindacalismo Colombiano* (Bogotá: Centro de Investigación y Accion Social, 1975).
Goodwin, Jeff, and Theda Skocpol, "Explaining Revolutions in the Contemporary Third World," *Politics and Society,* vol. 17, #4, September 1989, pp. 489–509.
Government of the United States, Accountability Office, *Drug Control: Aviation Program Safety Concerns in Colombia Are Being Addressed, but States' Planning and Budgeting Process Can Be Improved* (Washington, DC: GAO, July 2004).
———, Bureau of International Narcotics and Law Enforcement Affairs, "Resource Summary," February 2005, www.whitehousedrugpolicy.gov/publications/policy/06budget/state.pdf, viewed 2 June 2005.
———, "Colombian Narco-Terrorist Extradited to the United States," 31 December 2004, www.usinfo.state.gov/wh/archive/2005/Jan/03–115430.html, viewed 13 June 2005.
———, Department of Defense, *C⁴, Intelligence, Surveillance, and Reconnaissance* (Washington, DC: U.S. Department of Defense), 1996, www.dtic.mil/exec/adr96/chapt_27.html, viewed 14 October 1998.
———, Department of Defense, Defense Security Cooperation Agency, "Plan

Colombia," 2001, www.dsca.mil/programs.Comptroller/2001_FACTS/, viewed 6 October 2005.
———, Department of State, "Congressional Budget Justification for Foreign Operations, Fiscal Year 2005," 2004, www.state.gov/m/rm/rls/cbi, viewed 27 December 2005.
———, Department of State, "Fiscal Year 2007 Congressional Budget Justification—Western Hemisphere, Colombia," 13 February 2006, www.state .gov/documents/organization/60656.pdf, viewed 29 May 2006.
———, Department of State, "Foreign Military Training: Joint Report to Congress, Fiscal Year 2003–2004," June 2004, www.state/gov/t/pm/rls/rpt/fmtrpt/ 2004/34221.htm, viewed 3 October 2005.
———, Department of State, "International Narcotics and Law Enforcement—FY 2006 Budget Justification," April 2005, http://state.gov.p/inl/rls/rpt/cbj/ fy2006/511139.htm, viewed 3 May 2006.
———, Department of State, "President Bush, President Pastrana Discuss Trade, Terrorism, George W. Bush, President, Remarks with President of Colombia at the White House," 18 April 2002, www.state.gov/p/wha/rls/rm/9542.htm, viewed 18 March 2003.
———, Department of State, "Report on Certain Counternarcotics Activities in Colombia," April 2003, www.ciponline.org/colombia/03041401.htm, viewed 2 November 2003.
———, Department of State, "A Report to Congress on United States Policy towards Colombia and Other Related Issues," 3 February 2003, www.state.gov/p/ what/rls/rpt/17140.htm, viewed 3 July 2005.
———, Department of State, "State Department Overview of Terrorist Threat in the Americas," 29 April 2004, http://usinfo.state/gov/is/Archive/2004/Apr/29 -69799.html, viewed 23 May 2004.
———, Office on National Drug Control Policy, "The Andes: Institutionalizing Success," 11 May 2005, www.whitehousedrugpolicy.gov/news/testimony05/ 051105.coca.pdf, viewed 10 June 2005.
———, Office of National Drug Control Policy, "Technical Report for Price and Purity of Illicit Drugs," Appendix H, 2004, www.whitehousedrugpolicy.gov/ publications/price%5Ftech%Frpt/apph.pdf, viewed 10 June 2005.
Gramsci, Antonio, *Selections from the Prison Notebooks* (New York: International Publishers, 1971).
Guaqueta, A., "Change and Continuity in US-Colombian Relations and the War against Drugs," *Journal of Drug Issues*, vol. 35, #1, 2005, pp. 27–56.
Guevara, Che, *Guerrilla Warfare* (Lincoln: University of Nebraska Press, 1998).
Gunder Frank, Andre, *Latin America—Underdevelopment or Revolution* (New York: Monthly Review, 1969).
Gurr, Ted Robert, *Why Men Rebel* (Princeton, NJ: Princeton University Press, 1970).
Hall, Ninal, *Exploring Chaos: A Guide to the New Science of Disorder* (New York: Norton, 1994).
Harkentt, Richard, "The Risks of a Networked Military," *Orbis*, vol. 44, #1, 2000, pp. 127–143.

Hart, B. H. Liddell, *Strategy* (New York: Meridian, 1991).

Harvey, David, *The Condition of Postmodernity* (Cambridge: Blackwell, 1989).

Hayward, Keith, "The Globalization of Defense Industries," *Survival,* vol. 43, #2, 2002, pp. 115–132.

Hernández-Díaz, Jorge, *Reclamos de la Identidad: La Formación de las Organizaciones Indígenas en Oaxaca* (Oaxaca: Universidad Autónoma Benito Juárez de Oaxaca, 2001).

Herrera, Cecilia, "City as a Modernizing Paradigm," *Pedagogica historica,* vol. 49, #1, 2003, pp. 65–69.

Heyman, Josiah, "Ports of Entry as Nodes in the World System," *Identities,* vol. 11, #3, 2004, pp. 303–327.

Hobbes, Thomas, *Leviathan* (New York: Penguin, 1982).

Hobsbawn, Eric, *Rebeldes Primitivos* (Barcelona: Ediciones Ariel, 1980).

Holmes, Jennifer, Sheila Amin Gutiérrez de Piñeres, and Kevin Curtin, "Drugs, Violence, and Development in Colombia: A Department-Level Analysis," *Latin American Politics and Society,* vol. 48, #3, 2006, pp. 7–184.

Homer-Dixon, Thomas, *The Ingenuity Gap: Facing the Economic, Environmental, and Other Challenges of an Increasingly Complex and Unpredictable Future* (New York: Vintage, 2002).

Hoskin, Gary, and Gabriel Murrillo, "Colombia's Perpetual Quest for Peace," *Journal of Democracy,* vol. 12, #2, 2001, pp. 32–45.

Hussain, Imtiaz, "Of Mountains, Molehills and Mirages: Canadian and Mexican Central American Pursuits," *Revista Mexicana de Estudios Canadienses,* vol. 1, #7, 2004, pp. 1–10.

Ibugger, Lothar, "The Revolution in Military Affairs," Nato Parliamentary Assembly Paper, November 1998.

International Monetary Fund, Country Report, "Colombia,"#05/162, May 2005, www.imf.org/external/pubs/ft/scr/2005/cr05162.pdf, viewed 14 June 2005.

International Red Cross, *Estadísticas: Asistencia Humanitaria de Emergencia Colombia* (Geneva: International Red Cross, 2003).

Isaacson, Adam, "Optimism, Pessimism and Terrorism: The United States and Colombia in 2003," *Brown Journal of World Affairs,* vol. 10, #2, 2004, pp. 245–256.

Jaramamillo Uribe, Jaime, "Nación y Región en los Orígenes del Estado Nacional en Colombia," in *Travesías por la Historia,* ed. Frank Safford (Bogotá: Biblioteca Familiar Presidencia de la República, 1997), pp. 118–122.

Jomini, Antoine Henri de, *The Art of War* (London: Greenhill, 1996).

Jung, Courtney, "The Politics of Indigenous Identity: Neoliberalism, Cultural Rights, and the Mexican Zapatistas," *Social Research,* vol. 70, #2, 2003, pp. 430–449.

Kaplan, Robert, "The Coming Anarchy," *Atlantic Monthly,* vol. 273, #2, February 1994, pp. 44–76.

———, *The Ends of the Earth: A Journey at the Dawn of the 21st Century* (New York: Random House, 1996).

Karch, Steven B., *A Brief History of Cocaine* (New York: CRC Press, 1998).

Khaldun, Ibn, *An Arab Philosophy of History—Selections from the Prologomena* (Princeton, NJ: Darwin Press, 1987).

Kiel, Douglas, *Chaos Theory in the Social Sciences* (Detroit: University of Michigan Press, 1997).
Klare, Michael, *Resource Wars: The New Landscape of Global Conflict* (New York: Owl, 2003).
Klein, Harvey, *Colombia: Democracy under Assault* (Boulder, CO: Westview Press, 1995).
Kose, M. Ayhan, "How Has Nafta Affected the Mexican Economy?" IMF working paper, #WP/04/59, April 2004.
Krepinevich, Andrew F., "Calvary to Computer: The Patterns of Military Revolutions," *The National Interest,* #37, Fall 1994, pp. 30–42.
Kruijt, Dirk, and Kees Koonings, eds., *Societies of Fear: The Legacy of Civil War, Violence and Terror in Latin America* (London: Zed, 1999).
Lao Tzu, *Tao te Ching* (New York: Penguin, 1963).
Latham, Andrew, *Understanding the RMA: Braudelian Insights into the Transformation of Warfare* (Geneva: Programme for Strategic Studies and International Security Studies, 1999).
Leal Buitrago, Francisco, *La Seguridad Nacional a la Deriva: Del Frente Nacional a la Posguerra Fria* (Bogotá: Alphaomega, 2002).
———, "Las Utopías de la Paz," in *Los Laberintos de la Guerra: Utopías e Incertidumbres Sobre la Paz,* ed. Francisco Leal Buitrago (Bogotá: Tercer Mundo, 1999), pp. 109–170.
Leech, Gary, "Plan Petroleum in Putumayo," *NACLA Report on the Americas,* vol. 37, #6, 2004, pp. 8–11.
———, "Regime Change in Venezuela?" *NACLA Report on the Americas,* vol. 37, #6, May 2004, p. 22.
LeGrand, Catherine, *Close Encounters of Empire* (Durham, NC: Duke University Press, 1998).
———, "The Colombian Crisis in Historical Perspective," *Canadian Journal of Latin American and Caribbean Studies,* vol. 28, #55, 2003, pp 165–185.
———, *Frontier Expansion and Peasant Protest in Colombia* (Alburquerque: University of New Mexico Press, 1986).
Liga Mexicana por la Defensa de los Derechos Humanos, "Informe Sobre la Masacre del El Charco," 7 June 1999, www.derechos.org/limedddh/informes/charco.html, viewed 2 January 2003.
Lopera, Liliana, "When the Impossible in Juridical Terms Becomes Possible in Political Terms," Escuela Nacional Sindical, Derechos Humanos, *Report on the Violation to the Human Rights of the Colombian Union Workers, 2003,* Medellín, April 2004.
Lynch, Tony, "The Good Mercenary?" *Journal of Political Philosophy,* vol. 8, #2, 2000, pp. 133–153.
Lyon, David, "An Electronic Panopticon? A Sociological Critique of Surveillance Theory," *Sociological Review,* vol. 41, #1, 1993, pp. 653–678.
Machiavelli, Niccolo, *The Discourses* (New York: Penguin, 1970).
———, *The Prince* (New York: Mentor, 1980).
Mao Tse Tung, *On Guerrilla Warfare,* 1937, reprinted from original by Maoist Documentation Project, www.maoismorg/msw/vol16/mswv6-2g.htm.

Marcella, Gabriel, and Donald Schulz, *Colombia's Three Wars: US Strategy at the Crossroads* (Carlisle, PA: U.S. Army War College, Strategic Studies Institute, 1999).

Martz, John, "National Security and Politics: The Colombian-Venezuelan Border," *Journal of Interamerican Studies and World Affairs,* vol. 30, #4, Winter 1988–1989, pp. 117–138.

Marx, Karl, *The 18th Brumaire of Louis Bonaparte* (New York: International, 2004).

Mason, Ann, "Colombia's Democratic Security Agenda: Public Order in the Security Tripod," *Security Dialogue,* vol. 34, #4, 2003, pp. 391–410.

Mayor Mora, Alberto, *Etica, Trabajo y Productividad en Antioquia* (Bogotá: Tercer Mundo, 1984).

McDermott, Jeremy, "Farc and the Paramilitaries Take Over Colombia's Drug Trade," *Janes Intelligence Review,* 1 July 2004, www.janes.com, viewed 1 July 2004.

Medina Gallego, Carlos, ed., *ELN: Una historia contada a dos voces* (Bogotá: Rodríguez Quito Editores, 1996).

Metz, Steven, "The Next Twist of the RMA," *Parameters,* vol. 30, #3, Autumn 2000, pp. 40–53.

Meyer, Lorenzo, *Historia de la Revolución Mexicana: 1928–1934* (Mexico, DF: Editorial Joaquín Mortiz, 1998).

Miro, Ramon, "Organized Crime and Terrorist Activity in Mexico, 1999–2002," Federal Research Division, U.S. Library of Congress, Washington, DC, February 2003.

Mockler, Anthony, *The New Mercenaries* (London: Sidgwick and Jackson, 1985).

Moore, Barrington, Jr., *The Social Origins of Dictatorship and Democracy: Lord and Peasant in the Making of the Modern World* (Boston: Beacon Press, 1966).

Morgan, Patrick, "The Impact of the Revolution in Military Affairs," *Journal of Strategic Studies,* vol. 23, #1, 2000, pp. 132–163.

Musashi, Miyamoto, *The Book of Five Rings* (New York: Kodansha, 2001).

Mutis, Álvaro, *The Adventures and Misadventures of Maqroll* (New York: New York Review Books, 2002).

Napoleon, *How to Make War* (New York: Ediciones La Calavera, 1998).

Natividad Rosales, José, *Quien es Lucio Cabañas* (Mexico, DF: Posada, 1974).

Oficina del Alto Comisionado de las Naciones Unidas para los Refugiados (ACNUR), *La Población Desplazada en Bogotá* (Bogotá: ACNUR, 2003).

Organization for Economic Co-operation and Development, "Economic Survey of Mexico 2005," 12 September 2005, www.oecd.org, viewed 2 November 2005.

Orquist, Paul, *Violence, Conflict and Politics in Colombia* (New York: Academic Press, 1980).

Ortiz, Roman, "Insurgent Strategies in the Post-Cold War: The Case of the Revolutionary Armed Forces of Colombia," *Studies in Conflict and Terrorism,* vol. 25, 2002, pp. 127–143.

Ortiz-Lozano, L., et al., "Environmental Evaluation and Developmental Problems of the Mexican Coastal Zone," *Ocean and Coastal Management,* vol. 48, 2005, pp. 161–176.

Overmyer-Velasquez, Rebecca, "The Self-determination of Indigenous Peoples and the Limits of United Nations Advocacy in Guerrero, Mexico, 1998–2000," *Identities,* vol. 10, #9, 2003, pp. 10–22.

Owens, Mackubin Thomas, "Technology, the RMA and Future War," *Strategic Review,* vol. 26, #32, Spring 1998, pp. 63–70.

Parra, Francisco, *Oil Politics: A Modern History of Petroleum* (New York: Taurus, 2004).

Pastor, Manuel, Jr., and Carol Wise, "The Lost Sexenio," *Latin American Politics and Society,* vol. 47, #4, 2005, pp. 135–160.

Paulson, Joshua, "Rural Rebellion in Southern Mexico," *Nacla Report on the Americas,* vol. 33, #5, March 2000, pp. 4–5.

Pavón, David, "Cronología del EPR y del EPRI," *Revolución,* #25, June 2000, www.geocities.com/Pentagon/Bunker/5061/cron3.htm/200520, viewed 4 April 2004.

Pavón, David, and Maria Luisa Vega, "Cronología del Ejército Popular Revolucionario (EPR) y las Fuerzas Armadas Revolucionarias del Pueblo (FARP)," *Revolución,* #27, 2 May 2000, www.geocities.com/Pentagon/Bunker/5065/cron4.html, viewed 3 March 2002.

Paz, Octavio, *The Labyrinth of Solitude* (New York: Grove Press, 1985).

Pearce, Jenny, *Colombia: Inside the Labyrinth* (London: Latin American Bureau, 1990).

Pecaut, Daniel, "The Loss of Rights, the Meaning of Experience, and Social Connection: A Consideration of the Internally Displaced in Colombia," *International Journal of Politics, Culture and Society,* vol. 14, #1, 2000, pp. 89–106.

———, *Politica y Sindicalismo en Colombia* (Bogotá: La Carreta, 1973).

———, "Presente, Pasado y Futuro de la Violencia en Colombia," *Desarrollo Económico,* vol. 36, #144, January–March 1997, pp. 891–930.

Peñaranda, Ricardo, and Javier Guerrero, eds., *De Las Armas a La Política* (Bogotá: Tercer Mundo, 1999).

Peterson, V. Spike, *A Critical Rewriting of Global Political Economy* (London: Routledge, 2003).

Petkoff, Teodoro, *La Venezuela de Chavez: Una Segunda Opinion* (Caracas: Grijalbo, 2000).

Pius Llopart, Jordi, "Guerrero Ecologists Promote Drug Crop Alternatives," *Nacla Report on the Americas,* vol. 36, #2, September 2002, pp. 20–21.

Pizarro Leóngomez, Eduardo, *Insurgencia Sin Revolución: La Guerrilla en Colombia en una Perspectiva Comarada* (Bogotá: Tercer Mundo, 1996).

———, *Las Farc (1949–1966): De La Autodefensa a la Combinación de Todas Formas de Lucha* (Bogotá: Tercer Mundo, 1991).

———, *Una democracia asediada—balance y perspectivas del conflicto armado en Colombia* (Bogotá: Norma, 2004).

Portes, Alejandro, and Kelly Hoffman, "Latin American Class Structures: Their Composition and Change during the Neoliberal Era," *Latin American Research Review,* vol. 38, #1, 2003, pp. 1–35.

Puyana, Alicia, *Colombia: Economic Politica de las Expectivativas Petroleras* (Bogotá: Tercer Mundo, 2000).

Quedensley, Thomas, "The Commercial Satellite Multispectral Imagery Threat," *American Intelligence Journal,* vol. 21, #1, Spring 2002, pp. 33–36.

Ramírez, Socorro, *Colombia, Venezuela: Agenda comun para el siglo XXI* (Bogotá: IEPRI, 1999).

Ramírez Tobón, William, "Violencia, Guerra Civil, Contrato Social," in *Colombia: Cambio de Siglo; Balances y Perspectivas,* ed. William Ramírez Tobón (Bogotá: Planeta, 2000), pp. 21–68.

Rangel Suárez, Alfredo, *Guerra en el Fin de Siglo* (Bogotá: Tercer Mundo, 1998).

Rapoport, David, "Fear and Trembling: Terrorism in Three Religious Traditions," *The American Political Science Review,* vol. 78, #1, 1984, pp. 658–677.

———, *Guerros y Politicos* (Bogotá: Intermedio, 2003).

Rayner, Jeremy, "Labor Rights in Colombia," Henning Center for International Labor Relations, UC Berkeley, 2002, wwww.iir.berkeley.edu/henningcenter/gateway/Colombia.html, viewed 3 May 2003.

Rendón Alarcón, Jorge, *Sociedad y conflicto, en el estado de Guerrero, 1911–1995* (Mexico, DF: Plaza y Valdes, 2003).

Reyes, Alejandro, "Drug Trafficking and the Guerrilla Movement in Colombia," in *Drug Trafficking in the Americas,* ed. Bruce Bagley and William Walkers, III (Coral Gables: University of Miami, 1994), pp. 120–135.

Reyes, Eugenia, and Alvaro F. López, "Historia de la política agraria en Chiapas," *El Cotidiano,* #62, May-June 1994.

Richani, Nazih, *Systems of Violence: The Political Economy of War and Peace in Colombia* (Albany: SUNY, 2002).

Roberts, P., *The End of Oil: On the Edge of a Perilous New World* (New York: Houghton Mifflin, 2004).

Robin, Alexander, "Colombian Trade Union Leader Patricia Buritica Rallies Opposition to Plan Colombia," *Guild Notes,* vol. 24, #2, Summer 2000, p. 1.

Rochlin, James, "Canada, Narcotrafficking and Streams of Power," *Canadian Foreign Policy,* vol. 7, #1, Fall 1999, pp. 109–146.

———, *Discovering the Americas: The Evolution of Canadian Foreign Policy towards Latin America* (Vancouver, Canada: UBC Press, 1994), pp. 211–213.

———, *Redefining Mexican Security: Society, State and Region under Nafta* (Boulder, CO: Lynne Rienner, 1997).

———, *Vanguard Revolutionaries in Latin America: Peru, Colombia and Mexico* (Boulder, CO: Lynne Rienner, 2003).

Rodríguez Bautista, Nicholás (aka Gabino), "Ejército de Liberación Nacional: Una Historia de Vida," in *ELN: Una Historia Contada a Dos Voces,* ed. Carlos Medina Gallego (Bogotá: Rodriguez Quito Editores, 1996), pp. 27–168.

Romero, Mauricio, "Élites Regionales, Identidades y Paramilitares en el Sinú," in Peñaranda and Guerrero, *De Las Armas a la Politica,* pp. 175–218.

———, *Paramilitares y Autodefensas, 1982–2003* (Bogotá: IEPRI, 2003).

Rosado Duque, Juan Bernardo, Escuela Nacional Sindical, "Informe sobre la violación a los derechos humanos de los y las sindicalistas colombianos en el año 2005," mimeograph, Medellín, April 2006.

Ruelle, David, *Chance and Chaos* (Princeton, NJ: Princeton University Press, 1993).

Rulfo, Juan, *The Burning Plain and Other Stories* (Austin: University of Texas Press, 2003).

———, *Pedro Páramo* (New York: Grove, 1994).

Ryan, Terry, "Committing to a Future of ISR Supremacy," *American Intelligence Journal,* vol. 21, #1, Spring 2002, pp. 7–16.

Safford, Frank, and Marco Palacios, *Colombia: Fragmented Land, Divided Society* (New York: Oxford University Press, 2002).

Sarmiento Anzola, Libardo, "Debacle del Estado Social," in *Síntesis 2000: Anuario Social, Político y Económico de Colombia,* ed. Luis Restrepo Moreno (Bogotá: Instituto de Estudios Políticos y Relaciones Internacionales, Universidad Nacional de Colombia, 2000), pp. 27–39.

Scott, Peter Dale, *Drugs, Oil and War: The US in Afghanistan, Colombia and Indochina* (Lanham, MD: Rowman and Littlefield, 2003).

Segura Escobar, Nora, "Colombia: A New Century, an Old War, and More Internal Displacement," *International Journal of Politics, Culture and Society,* vol. 14, #1, 2000, pp. 107–119.

Shifter, Michael, "Breakdown in the Andes," *Foreign Affairs,* vol. 83, #5, 2004, pp 126–138.

Shimeall, Timothy, et al., "Countering Cyber War," *Nato Review,* vol. 49, #4, Winter 2001, pp. 16–18.

Singer, P. W., *Corporate Warriors: The Rise of the Privatized Military Industry* (Ithaca, NY: Cornell University Press, 2003).

Skocpol, Theda, *States and Revolutions* (New York: Cambridge University Press, 1979).

Smil, Vaclav, *Energy at the Crossroads: Global Perspectives and Uncertainties* (Boston: MIT Press, 2003).

Smith, Eugene, "The New Condotieri and US Policy: The Privatization of Conflict and Its Implications," *Parameters,* vol. 32, #4, Winter 2002–2003, pp. 102–125.

Sowell, David, *The Early Colombian Labor Movement: Artisans and Politics in Bogotá, 1882–1919* (Philadelphia: Temple University Press, 1992).

Stokes, Doug, "Why the End of the Cold War Doesn't Matter: The US War on Terror in Colombia," *Review of International Studies,* vol. 29, #4, 2003, pp. 569–581.

Sun Tzu, *The Art of War* (Boston: Shambhala, 1991).

Sun Tzu II, *The Lost Art of War* (San Francisco: HarperCollins, 1996).

Szafranski, Richard, "A Theory of Information Warfare: Preparing for 2020," *Airpower Journal,* vol. 9, #1, Spring 1995, pp. 56–65.

Tanenbaum, Frank, *Peace by Revolution* (New York: Columbia University Press, 1933).

Tarnoff, Curt, and Larry Nowels, Congressional Research Service, *CRS Report for Congress: Foreign Aid; An Introductory Overview of US Programs and Policy* (Washington, DC: Library of Congress, 2004), pp. 1–35.

Thucydides, *A History of the Peloponnesian War,* ed. Robert Strassler (New York: Free Press, 1996).

Tilly, Charles, "War Making and State Making as Organized Crime," in *Bringing the State Back In,* ed. Peter Evans (Cambridge: Cambridge University Press, 1985), pp. 169–187.

Toffler, Alvin, and Heidi Toffler, *War and Anti-War: Survival at the Dawn of the 21st Century* (New York: Little, Brown, 1993).

Tomes, Robert, "Relearning Counterinsurgency Warfare," *Parameters,* vol. 19, #1, 2004, pp. 16–28.

United Kingdom Parliament, "Drugs," 15 February 2006, www.publications.parliament.uk/pa/cm200506/cmhansard/cm060215w27.htm, viewed 28 May 2006.

United Nations Departamento Población, Equipo Coordinador: Informe Nacional de Desarrollo Humano, *Andes 2020 y el Conflicto, callejón con salida: Una Mirada Comparativa* (Bogotá: UN Departamento Población, 2004).

United Nations Development Program, "Colombia 2005," www.hdr.undp.org/statistics/data/countries.cfm?c+COL, viewed 3 May 2006.

United Nations Economic Commission for Latin America and the Caribbean (ECLAC), *Notes,* #43, November 2005.

United Nations Educational, Scientific, and Cultural Organization (UNESCO), "Displacement, Development, and Modernity in the Colombian Pacific," pamphlet (London: UNESCO, 2003).

Urrutia, Miguel, *The Development of the Colombian Labor Movement* (New Haven, CT: Yale University Press, 1969).

Vallejo, Fernando, *Our Lady of the Assassins* (London: Serpent's Tale, 2001).

Van Cott, Donna, "Andean Indigenous Movements and Constitutional Transformation," *Latin American Perspectives,* vol. 30, #1, January 2003, pp. 49–69.

Vargas Mesa, Ricardo, *Fumigacion y Conflicto: Políticas antidrogas y deslegitimación del Estado en Colombia* (Bogotá: Tercer Mundo, 1999).

Vico, Giambattista, *The New Science of Giambattista Vico* (Ithaca, NY: Cornell University Press, 1976).

Villegas, Jorge, *Petroleo Colombiano, Ganancia Gringa* (Bogotá: Ancora, 1999).

Virilio, Paul, *Speed and Politics* (New York: Semiotext(e), 1977).

Virilio, Paul, and Sylvere Lotringer, *Pure War* (New York: Semiotext(e), 1997).

Walker, R.B.J., *Inside/Outside: International Relations as Political Theory* (Cambridge: Cambridge University Press, 1993).

Washington Office on Latin America, *Colombia Monitor,* vol. 2, #1, May 2003.

Weber, Max, *Economica y Sociedad, vol. 2* (Mexico City: Fondo de Cultura Económica, 1977).

Wickham-Crowley, Timothy, *Guerrillas and Revolutions in Latin America* (Princeton, NJ: Princeton University Press, 1992).

Wilpert, Gregory, "Community Airwaves in Venezuela," *NACLA Report on the Americas,* vol. 37, #4, January 2004, pp. 34–40.

Woloski, Andrea, "Clash of Class," *Harvard International Review,* vol. 26, #4, Winter 2005, pp. 1–7.

Womack, John, *Zapata y la Revolución Mexicana* (Mexico, DF: Siglo Veintiuno Editores, 1999).

Wrighte, Mark, "The Real Mexican Terrorists: A Group Profile of the Popular Revolutionary Army," *Studies in Conflict and Terrorism,* vol. 25, #1, May 2002, pp. 207–225.

Wriston, Walter, "The Twilight of Sovereignty," *Fletcher Forum,* vol. 17, #2, Summer 1993, pp. 117–130.

Youngers, Coletta, and Eileen Rosen, eds., *Drugs and Democracy in Latin America: The Impact of US Policy* (Boulder, CO: Lynne Rienner, 2004).

Zahniser, Steven, and William Coyle, "US Mexico Corn Trade during the Nafta Era," U.S. Department of Agriculture, FDS-04D-01, May 2004.

Zuluaga Nieto, Jaime, "De Guerrillas a Movimientos Políticos," in Peñaranda and Guerrero, *De Las Armas a La Política,* pp. 1–74.

Index